Becoming a Man Through Military Service

Published by:
Eastern Offset Publishing Co.
PO Box 1091, Atlantic Beach, NC 28512

Author: Charles S. Manooch, III
Graphics & Design: Rebecca McMillan
Photos: Fort Jackson 1955 yearbook, Newsfoto Publishing Co.

Other Books by the Author:
Spring Comes to the Roanoke
Fisherman's Guide to the Fishes of the Southeastern United States
Growing up an Old Raleigh Boy

Library of Congress Control Number: 2010927192
ISBN: 978-0-9659506-6-4

First Printing 2010

Printed in the USA by:
Eastern Offset Printing Co.
Atlantic Beach, NC

Becoming a Man
Through Military Service

Charles S. Manooch, III, Ph.D.

2010

This book is dedicated to the memory of Emerson Atkinson who died way too soon in 2007. I had known Emerson since we were five-years old, and we maintained that friendship for sixty years. Emerson, and his father, kept in touch with my parents and me long after I had moved from Raleigh. Emerson completed the ROTC program at North Carolina State University, and then served as an officer in the Army. He and I discussed this book while it was being drafted, and he seemed keenly interested in it.

AUTHOR'S NOTE

I don't believe I could write a novel. It's not because I don't like to write, and those that have been around me know that I like to tell stories. However, I couldn't write about something that's pure fantasy. My long-time friend, Tommy Green asked me one day, "Chuck, why don't you write one of those mysteries like some of the popular authors do? They put out a new book every other month, the books are best sellers, and they make a lot of money. You could take a subject like a detective trying to solve a string of murders out in California and go with it."

I thought about it for a minute. " Because I'm not a detective, and have never been interested in solving a crime. Heck, I've never even been to California—although I did fly over it once."

Don't get me wrong. I greatly admire those who create novels, and I receive much pleasure from reading their books. It ranks up there as being one of my favorite pastimes. However, for me to compose a non-technical text I must have *lived* it—immersing myself in the sights, sounds, and smells of each scene that I attempt to portray. In that way I can use factual events and expand upon them in my own storytelling fashion. I try to do this in *Becoming a Man Through Military Service* as I did in *Growing Up An Old Raleigh Boy.* You'll be the judge of whether I was able to do that here successfully.

This book begins where my Raleigh book left off. Following kindergarten, elementary school, high school, and one year of college, I was rapidly approaching what seemed to be a dead end in my life and really didn't know what to do. After one summer at home I enlisted in the North Carolina National Guard on my father's advice—as a chance to receive discipline, and, hopefully, find some direction for my future. That, in a nutshell, is what this book is about—my active duty at Fort Jackson, SC from November 15, 1962 through May 15, 1963.

When my children and grandchildren read this book they will be surprised to find me using some coarse words. I was—and wanted to be—part of a group of young men who shared a life. We developed common actions, thoughts, and speech patterns. We were trained to be a unit, and that training was successful by any measurement. Some habits, like the use of verbal expressions offensive to other audiences, fade away quickly. But, the value of that training and life experiences endure.

I very much appreciate the editorial efforts that my good friend Gene Hafer of Raleigh and Ashe County, dedicated to assist me prepare drafts of the book. I also wish to thank Beki McMillan, Eastern Offset Printing Company, Atlantic Beach, NC for creating the layout and design, and for printing. Newsfoto Publishing Company of San Angelo, Texas should be acknowledged for the photographs which were taken at Fort Jackson, but not at the same time that I was there.

CHAPTER ONE

Tiny beads of sweat glistened in the early morning sun as they grew in size and slowly meandered down the poor fellow's nose. They were not the result of excessive labor or exercise, but rather the stress brought on by nausea. The man was middle-aged and overweight, and as I glanced over at him I could see ever increasing discomfort as his Adam's apple bobbed up and down and his eyes shot this way and that. It was obvious that he was at war with his gorge. "Sir, can I help you?" I said from across the aisle as he struggled to extract himself from his seat.

"Hell no, son. This I have to do by myself." He slowly got up and squeezed his way into the small bathroom located directly behind him. Minutes later he emerged, followed by the unpleasant odors of throw-up, stale urine, and sickening sweet disinfectant. This mingled with the pervasive smell of diesel exhaust as he and I sat in the back of a Greyhound bus, accompanied by about thirty other passengers as we traveled from Raleigh to Columbia, South Carolina.

Eighteen-and nineteen-year-old boys were disproportionately present as passengers on the bus. Hundreds of Army recruits across the southeastern United States were on their way to Basic Training at Fort Jackson. I was one of them. We were being packaged crosswise in time as we left one world and entered another. Young faces, incessant nervous chatter, and small overnight bags—typical of high school athletes—easily identified us. Some had their hair styled in crew cuts. Others had "ducktails" that seemed glossy, but were really greasy.

I could tell that some of the recruits were Yankees. You can tell by the way they talk, and also by their appearance. Yankees just look different, and are as strange looking to my biased eyes as folks from other countries. I have often—since then—sat in an airport and tried to assign a country of origin to a particular individual. East Asians, Indians, and Pakistanis are easy. Those from Mediterranean regions are typically swarthy, while Scandinavians are huge. Those men, women, children, and even babies, are large. Most Yankees are as easy to differentiate. For example, many of the older men have unusually large tufts of hair sticking out of their ears. You can look for that even before you hear their often loud and opinionated voices.

Okay. I know you can say Southerners are just as readily stereotyped. They're "rednecks," "crackers," or "hicks" and they sure as hell talk funny. You're dead right. I'll take that in stride. And, I have to admit that I notice the Southerners' use of the overly cute "thank yew" more than I used to.

The trip to Columbia from Raleigh should take about three and a half or four hours to travel the two hundred and thirty or so miles. That's in a car and not in a bus that makes frequent stops along the way. So it took us longer—a little over five and a half hours. The bus must have followed U.S. Highway 1 as it went through the towns of Southern Pines and Hamlet in North Carolina and Cheraw and Camden in South Carolina. We stopped briefly in each and I glanced up from my nap as a few passengers were exchanged at the stations. However, you can bet that we Army recruits were all *wide-awake* when the bus delivered us to our Columbia destination.

The waiting area was relatively quiet at one o'clock in the afternoon, and I sat there by myself a few minutes watching people. Some of the other recruits sort of rambled around—not knowing what to do—as if waiting for someone to step forward and take charge. There was one thing for certain, and that was my stomach was nervously tied in knots and was taking charge of me—and I desperately needed to find the men's room. It was there in a locked stall that I witnessed a sight, which has unfortunately remained forever in my mind. It wasn't all the stuff and pictures scratched on the door and walls. I had seen all that vulgarity before. That timeless philosophical passage: "Here I sit all broken hearted; paid a quarter and only farted"—had been repeated one time or other by most twelve-year-old boys in Raleigh and everywhere else for that matter. Besides, I had only paid a dime. Rather, it was the giggling and scratching sounds that were coming from the stall next to mine. I was aware of someone or something looming over me, and I looked up and thought: *What in the world is a chicken neck doing hanging from that partition?* To my utter amazement what I was looking at was no chicken neck, but rather the wrinkled up scrotum of a naked old man. I couldn't see the other part because he had hold of that. My first thought was that he looked like a human buzzard perched on a limb, and I wasn't about to wait for him to pounce. I pulled up my pants, ran out the door, and told a policeman what I had seen. He told me the man was a frequent visitor to the bus station's men's room who didn't hurt anyone, but regularly caused a disturbance. I could attest to that, and wondered why he seemed to be a fixture at the terminal and wasn't put away someplace where he could have gotten some help. There are certainly places back home where people like that can be taken care of. South Carolina would seem to be no different.

All the recruits had gathered in one area of the waiting room and looked anxious to get this next step—the trip to the base—over with.

"Good God, what a pretty bunch! All you lovelies get up NOW and come over here!" The man glaring at us was a Staff Sergeant dressed head to boot in starched khaki. His boots were black and so shiny the toes sparkled in the sun. He wore a cap that was angled slightly to one side and looked like an inverted boat—and which we later learned was identified by a vulgar term. A sky blue scarf was tucked in the shirt at his throat, and braids of the same color were around his right shoulder and under his upper arm. This color designated his MOS (Military Occupational Specialty) as that of an infantryman.

"When I call out your name and serial number, line up over there!" Names and numbers were shouted out and then he got to mine. "Manooch—Charles S. the third. Now would you believe that somewhere out there are two others just like him? Okay, Charles the third, your number is NG24970479. The 'NG' stands for 'No Good'—that's National Guard. Some of you are like Charles the third here and are 'six month wonders' or 'weekend warriors.' You other fellows that are 'RA's or 'US's are in here for a longer stint, and most of you *want* to be in the United States Army!"

I didn't know it then, but Fort Jackson's most decorated war hero was Sergeant Richmond Hobson Hilton, a South Carolinian who served with his state's National Guard in World War I. Hilton Field at Fort Jackson is named in his honor. That bit of information about the National Guard would've come into great use while I was stationed there.

After we were all accounted for, we were herded into another commercial bus that carried us out to the Base. The angry Sergeant came with us.

"Okay fellows. You're now entering the gates of your new home—Fort Jackson." The bus driver was smiling as he looked back at us in his rear view mirror, and had obviously assumed the role of a tour guide. He was—without question—very familiar with the base and its history, and seemed eager to share his knowledge with his captive, yet very attentive, audience.

"This land was set aside way back in 1917, before the First Great War, and was then called 'Camp Jackson' in honor of Andrew Jackson, Major General of the Army and seventh President of the United States." We all stared out the windows

at the low sandy ridges dotted with "turkey oaks" and small pine trees. It resembled all the other military bases in the Carolinas that I have seen since. The type of land that's not useful for much of anything except serving as a training base. Nothing much would grow there with the exception of the small trees, and they had struggled through decades to attain the pitiful sizes that they were.

"Soldiers that train here in the winter claim that this is the only place they know of where you can stand in snow and have sand blow in your eyes at the same time." *Guess we'll get the chance to see if that's true*, I thought, as I closed my eyes and tried to escape this time and place.

As we drove onto the military reservation that day I had the feeling something was out of order—that I was in the wrong place at the wrong time. I had that feeling not long ago while sitting in a small country church in western North Carolina. A young man had died suddenly at the age of twenty-nine and had left a pretty young wife, a two-year old daughter, and his parents (who were my age) to mourn. Most in the congregation were far too young to be attending a funeral. We sat there listening to music played by a fiddle and harp, and the notes of old time, simple hymns drifted over the congregation. They flowed out the open windows and faded away. But, for some of us older folks brought back fond memories of those resting under large trees outside in the small cemetery. It was as if we were all out of place and temporarily caught up in this peaceful, restful period. I was in a daze.

"That's Tank Hill up ahead there," the driver said as he nodded his head forward.

"What's Tank Hill?" one of the fellows inquired.

"You-all will get to know it soon enough after you've run up and down it a few times and wished you were back home with your mamas," the Sergeant snorted. "In about five minutes we are going to unload at the U.S. Army's Reception Station. Then, you will truly be under my loving care for the rest of today and tonight."

"Good grief. That sounds like *fun*," moaned a pimpled-face boy who sat next to me. He had been nervously kneading his seat's armrest ever since we left the Columbia bus station. He was from Loris, a little town in South Carolina, and I doubt if he'd ever been more than ten miles from home. "I wonder if it's too late to get out of all this?" he said as he looked at me and managed a weak smile.

"Unfortunately, yes," I replied.

CHAPTER TWO

"WELCOME TO THE U.S. ARMY.

VICTORY STARTS HERE!" The sign greeted approximately three hundred recruits each day as they arrived at the Reception Station. It had initially been called the Personnel Center back in 1950 when inductees were trained for the Korean Conflict. In 1960 the Center was reopened as the Reception Station, and has remained an integral component of the U. S. Third Army. *(We soon learned what the Third Army shoulder patch looked like. It had a white "A" on a blue field, surrounded by a red circle. The patch is still in use today).* The Station would be home for inductees for three to five days, until each soldier had been "processed" before being assigned to a Basic Training Unit. When told that we were to be "processed" I couldn't help but think of other types of processing. The image of chickens that were suspended by their necks on a conveyor apparatus, and then plucked, gutted, and beheaded came to mind. Or, as my grandmother Loulie did in Raleigh on some Sundays after church, by grabbing a chicken, lopping off its head and throwing it under a number three washtub to keep it from running off. That's how she did her *processing.* That didn't happen to us—but we came close.

Unfortunately, when we entered active military service that day it was long before the tragic mishap at Paris Island, SC that involved Marine recruits during a training exercise. Several young men drowned one night when they were force- marched, heavily laden with backpacks, into a swamp. The public outcry was tremendous, and resulted in changes in the ways military recruits were disciplined at bases involving all branches of the service throughout the United States. Recruits are no longer to be so physically abused. On November 15, 1962, when we lined up in our first pitiful formation, we could be. Also, we quickly learned the cadre were not too gentle-spoken when they wanted to get our attention.

"Ladies! Ladies! I'm Sergeant Treadwell Hoffman. I've been in this man's Army for ten years. My daddy was in the Army for twenty and he fought in Europe in the Second World War. I served in that stinking hellhole known as Korea, and I've killed more than a few chinks." The angry man who had herded us around thus far had finally given up his name. "Now, all of us cadre at this base are responsible to see that each of you is trained and equipped for combat to prevent *you* from being killed. The first order of business today is that you are going inside—where each of you will hand deliver your orders. After that, we are going over to your barracks to get you quickly settled in, and then you-all are going to get your hair styled—just like Uncle Sam wants you to look."

"Sergeant Hoffman, sir. Do we get to tell them how much we'd like it cut?"

"Sir?! Sir?! Do I look like a fucking officer to you boy? Let's get something straight right now! You shit heads are all E-1s and that makes you as sorry as they come. I'm an E-7, a hard working Noncommissioned Officer, also known as an 'NCO' or 'Noncom.' If I had wanted to look pretty and strut around doing absolutely nothing then I would be an officer. And to get back to your question, dumb ass, you will not get to TELL *anything* to *anybody* as long as you're on this Base. Do you get that?"

"Yes sir! I mean, yes *Sergeant*!"

Our barracks were two-story wooden buildings painted off-white. Come to think of it, I soon learned that anything that was wood and painted in the Army was white, brown, light green,

or dark green, and I have no earthly idea what a Sergeant would have done with a bucket of yellow, blue, or red paint. That's probably a good thing, since a fundamental element of combat is to keep the enemy from seeing you, and bright colors would certainly take away from any attempt to camouflage.

Along that same line of reasoning, it didn't take me and some of the other fellows long to wonder why the Army would send us to face an enemy while wearing olive drab fatigues, and at the same time have high-neck white t-shirts and our names printed in black on a white background across the right side of our chests. Now, you talk about two perfectly identified targets!

All that changed in a hurry a few years later when U.S. troops fought in the jungles of Viet Nam.

The barracks were old, dating back to 1941, and as recently as 1960 there were more than 2,600 on Base. Each was large, elongated, and in a very general way, reminded me of a big South Carolina tobacco barn—except the barracks had a wooden ladder fire escape next to a door in the end of the buildings. North Carolina had its own tobacco barns of course, but there was something about the overhangs of the roofs between the first and second floors of the barracks that to me resembled old South Carolina tobacco barns.

Inside each barracks there were two large open bays—one was located on the first floor, and one was on the second. Rows of double bunks had their foot ends facing the bays' center walkways. Footlockers were at the ends of the bunks. A series of wall lockers, like those in high schools, were about half way down each row of bunks. We were instructed to temporarily store the personal items we had brought with us in our footlockers. Our personal gear was very limited when we arrived on Base, and underscored the fact that we were not going to be civilians for more than a day. We had been instructed to bring the following items: One set of casual clothing (no shorts), three sets of underwear (white), one pair of white calf-length socks, one pair of athletic shoes, one athletic supporter (men only in those days), one small gym bag, one disposable razor, shaving cream, one six-inch black comb, one bar of soap, one washcloth, one towel, non-aerosol antiperspirant, and five dollars cash.

After a very brief respite in the barracks we were herded back into the street. Our next stop was relatively close by and within easy walking distance. Recruits at Fort Jackson soon learned that everything on the military reservation was supposedly within *walking* distance. This was the infantry after all.

The barbershop, if you could call it that, was by far the largest I had ever seen. There must have been fifteen barbers and all of them were cutting hair at the same time. My first impression was of the hair. Hair was everywhere. There were piles of brown hair, black hair, and blonde hair. Some of it was kinky, some straight, some curly, some fine, and some greasy. Torrents of hair began falling off heads with stunned faces held up by skinny necks. It was literally all over in a matter of seconds. I had never seen anything like it. Only the Negro boys, and there were just a few of them, looked about the same when they came out as they did when they went in. Their hair was relatively short and tight so the "before" and "after" contrast was not as drastic. The stunned white guys in particular that stumbled out of there looked pathetic. I don't believe I looked quite so bad. I had a short thick neck and I weighed two hundred and three pounds. (Although I saw myself as fit, I was later assigned to the "fat boys table" in the mess hall along with some of the *truly* overweight guys.) The other skinned recruits looked like survivors of Auschwitz after their "shearing," and it was a stretch of the imagination to visualize these fellows as becoming rough and ready, real soldiers in the U.S. Army. Hell, you wouldn't have even selected one to play on your side in a touch football game.

Outside, we once again came together in what could casually be called a formation. "I'm getting hungry," I whispered to the guys close to me as we waited for further instructions, which I hoped would include a trip to the mess hall.

"You look like you could always be hungry. But it doesn't look like we'll be first in line." A black recruit named McCoy looked at me and then pointed to a group of soldiers in uniform that were marching across the street. McCoy was from Baltimore, and as with most of the other guys at Fort Jackson I only remember his last name. That was what we were called by, and that's what was printed on our uniform shirts, which were later issued to us. Standing there, I looked around for the boy from Loris who was on the bus with me when we left Columbia. I didn't see him, and in fact I never saw him again. I believe that some times folks just disappear for no apparent reason, particularly if they're off somewhere far from home, and have nobody close by to look after them. They're just gone.

The inductees across the street were actually marching in formation, and had obviously been at the Reception Station at least a day longer than we had. From that point on, each unit was evaluated with a "time on Base" criterion and only a day or two could establish some seniority. Recruits that arrived two days earlier looked down on day-one inductees.

The "old" troops' voices responded to their drill Sergeant's commands, and that ever-familiar infantryman's song filled the air:

"Hup, two, three, four! Hup, two, three, four!
Delayed cadence; delayed cadence; delayed cadence
c-c-o-o-u-u-n-n-t-t:
One (*two, three, four*)!
Two (*two, three, four*)!
Three (*two, three, four*)!

Four (*two, three, four*)!
One (pause)! Two (pause)!
Three (pause)! Four (pause)!
One, two, three, four!
One, two, three, four!"

"I guess it won't be long before we can do that too," said Toby Lee, the boy from Salisbury who had been right with me since we left Raleigh. Another fellow with the last name Lee was with me during training. He was from Clinton, NC and was like more than a few guys I met who was unable to read or write.

Toby Lee looked awful. He was naturally as white as a sheet, and his long neck and small head, now highlighted by his new "hair style", made him look like a turkey. A pimpled face turkey that was wet.

"I guess so," Mc Coy said. "But I'm like Ma-nuke here. I want something to eat before I even think about marching like that. It's been a long trip. Shit! I left Baltimore yesterday."

The wonderful smells of cooked vegetables, meat, and fresh bread greeted us as we lined up outside the mess hall. The place was like a big cafeteria where two hundred inductees at a time went through a serving line, and then ate while seated at long tables. Since this was our first experience at military dining, we were given special treatment. We were served by Sergeants who stood holding either a large aluminum fork or spoon, behind whatever dish each was responsible for. The presence of these NCOs was not in our *honor*, but to ensure that we learned *how* to be served. We carried compartmentalized aluminum trays and were instructed to hold them out from us, and *immediately adjacent to the tray or pot* containing the food. If the server plopped the food off his spoon or fork and it missed your tray, you were told to move on and you were not served that item. It was a mess. Food was all over the floor and along the rails where trays had been held in the "wrong" way, and we had to be careful not to slide in all the sticky goop as we shuffled along.

McCoy, Lee, and I managed to catch all of our food, and went to a table and joined five or six other guys. We gobbled down country-style steak, mashed potatoes, mixed green peas and carrots, and topped it off with milk or unsweetened ice tea. We had lemon meringue pie for dessert. The food was good, hot, and more than satisfied our hunger.

"That weren't bad," Toby said as he wiped his mouth with his shirtsleeve, and got up to go back outside. We joined him and our chairs loudly scraped across the dark green tiles of the floor.

"I need about ten good volunteers to step forward and show their Army spirit," said a large man wearing a white apron and hat. "Don't be shy. Step on up here."

"The guys in my Guard unit back home told me to *never volunteer* for anything while in the Army," I whispered as I looked down at my shoes and tried as hard as I could to be as inconspicuous as possible. The Mess Steward was now walking down a row of tables and was coming dangerously close to where we were grouped up.

"You, you, you and you," he kept repeating until ten inductees had been singled out. You guessed it. I was one of them. Lee and McCoy slipped back out to the street and left me standing there to meet my fate. I have often thought what the probability was of selecting me as one of ten among two hundred, which is complicated by the fact that another group of inductees had eaten supper before us. I have no idea how many were served, and our ten was representative of both groups. It really makes no difference. In any event, I had received my sentence.

When the mess hall had been cleared, the Sergeant called us "chosen few" back into the kitchen. It was by far the largest kitchen that I had been in. Everything was stainless steel—a walk-in refrigerator or cooler, several stoves with ovens and big sinks. "Okay guys. You are now on KP. That stands for Kitchen Police, and it will certainly not be the last time you have this wonderful opportunity while you're on Base. All but one of you will work inside doing exactly what you're told, and that one—that very fortunate fellow—will be the 'outside man.'" Thank God I wasn't chosen for that. The outside man was responsible for emptying all the garbage cans and then scrubbing them out with soap and hot water. That was bad enough. However, his real, real dirty job was to clean out the grease trap, which was located in a drain several feet beneath a concrete cover. Nine of us stood there looking out the back door as the "outside man" lifted the cover to the pit and slowly disappeared into what was a foul-smelling hole. It was so cold outside that steam was rising from the entire cleaning area.

"Damn! What a hell of a job," one of the nameless inductees said, shaking his head as we turned back in the kitchen to get our instructions. We dumped leftover food into galvanized metal garbage cans and two of us worked together at a time to carry them outside. Trays, pots, pans, serving utensils, glasses, cups, knives, forks, and spoons were cleaned in large dishwashers that sat on top of stainless steel work surfaces next to the sinks. "What's that thing, Sergeant?" one of the KPs asked as he pointed at what looked like a big stainless steel barrel with holes in it.

"That's an electric potato peeler," the Sergeant replied. "But you'll never use one while you're here at Fort Jackson. Trainees peel all their potatoes by hand."

After we'd cleaned the kitchen spotless, we had to mop the floors and then wax and shine them with big electric buffers that were difficult to operate without losing control of them. They reminded me of bumper cars at the North Carolina State Fair—except these Army buffers didn't have steering wheels. Two hours later the mess hall and kitchen looked spotless, and we proudly gazed around and eagerly waited for our orders to return to barracks. We were all very tired from the long day.

"One more job to do and then I'll release you." The Mess Sergeant opened a storage closet and placed several cans of paint and some brushes on a table. "I need to have a couple of doors and window sills repainted, and they need to dry before we serve breakfast tomorrow."

"Jesus Lord. Hasn't he gotten enough out of us by now?" I said as I picked up a brush and started to paint a windowsill. He hadn't, and we stayed there until ten o'clock, scrubbed the grease and paint off our hands, and stumbled back to our barracks. It was amazingly quiet. And on the short walk back I could smell coal smoke in the chilly night air—something that will always remind me of Fort Jackson.

I thought back to my childhood when Butch Royster, along with his two brothers and his mother and dad, lived in a house in Raleigh that was heated by a coal-burning furnace. The house is located at the corner of Fairview and Oberlin roads next to Fire Station Number Six. Butch's rise to fame at the time was derived from the fact that he could forcefully swallow air, and then burp the entire first verse of *Dixie*. Emerson Atkinson, Howdy Manning, Frank Crowley, and I used to visit the Roysters' when we were young boys to play—and to be serenaded by Butch. The coal smoke at the Base and at Butch's house sure smelled the same.

The barracks were dark when I got there. Inside, I got a quick shower before falling into my bunk for sleep that I feared wasn't going to last nearly long enough. The lights were already out, and I lay there a few minutes before I drifted off—listening to the snoring coming from my new comrades.

CHAPTER THREE

"Let go of your cocks, and on with your socks!" It took me a moment or two to figure out just where in the heck I was, and what in the world was that mad man shouting about?

My God, I thought I was dreaming. But I'm really here. I am in the Army—and this is Fort Jackson. It was like waking up to the truth, a truth you're not prepared to face—like someone close to you being extremely sick. I have a doctor friend who refers to *real* sick patients as being "low sick." When he says, "low sick" then you'd better get things in order and call Hospice.

That's almost the kind of dread I had lying there as the Sergeant shouted orders to us, and then started blowing his damn whistle. Clothes were thrown on and the barracks emptied in only a few minutes. Hardly any words were spoken as we spilled out into the street as a record played "Reveille"—loudly signifying the call of the first formation of the day. Our day. The first of many we were to begin while on active duty.

"AT-TEN-SHUN! Fall in! Come on you sissies. Try to line up and stand at the proper position! It's not hard! Look straight ahead. Arms down by your sides. Feet at a forty-five degree angle!" Each barracks had four drill Sergeants who moved through our ranks giving individual instructions in the pitch dark.

"Left face! That means turn to your left. Now! When I order you to march, you will start off on your left foot and move at a four count pace. Like this: 'One, two, three, four; one, two, three, four.' Left feet are odd numbers; right feet are even. I want you to stay in step. Now, forward march!" The formation sort of snaked its way towards the mess hall, and probably would've looked much worse if seen in broad daylight.

The breakfast was first rate—with eggs, bacon, cheese grits, home fries, and toast. The coffee was good too, although it was rumored that the cooks used t-shirts to strain the coffee from the grounds. Some of the fellows had never seen grits before so it was amusing watching them trying to decide what to put on them. We were not allowed to sit there long when we had finished eating, and this time none of the guys in my group were singled out for KP.

Sunlight bathed the Reception Station, and was shining

in our eyes as we marched over to a building where shots were to be administered to each inductee. As a ten-year-old, I had been afraid of getting my shot as school let out for summer. I would have been scared to death if I had known what was in store for me this morning. Inside, we were instructed to form lines that meandered through different medical stations. Individual heights and weights were recorded; doctors performed quick inspections of our mouths, ears, noses, balls, and rear ends. Dentists examined our teeth. Next, we proceeded to where we were to be inoculated. The Army was experimenting with a new type of air gun that was used to deliver the medications. Sadly for us, the gun operators had received only a little "on the job training" and were inductees like us who had arrived on base only two days before.

"I'm not going to let some redneck from Georgia give me a damn shot," McCoy snorted as we slowly moved ahead.

"You don't seem to have a choice," I replied as we all peered over or around the shoulders of the fellows in front of us to see what was happening up ahead. Each inductee was required to stand between two guys with air guns pointed at both shoulders. Two shots were delivered at the same time, and then the trainee moved to the next station. The tendency was to recoil from one injection and then literally bump into the next one on the opposite side. This was repeated until a total of thirteen shots had been administered.

We were still wearing our civilian clothes that morning, so it wasn't unexpected when our company was paraded into a large quartermaster's storage and distribution building where we were to receive our military clothing and supplies. By this time it was becoming obvious that most of us would be together during Basic Training. Besides McCoy and Lee, there was a tough-acting boy named Mills from Monks Corner, SC, and a large, soft-spoken fellow named Johnson from Archer Lodge, NC, and we had already formed a small group. It's human nature to be gregarious and come together during times of trouble or uncertainty, and these sure were troubled times, or so we imagined. Mills, like Johnson, was a white fellow, but he talked different and looked a bit mixed up—if you understand what I mean. He'd say "dem" and "dose" when referring to somebody—but so did a couple of other trainees from Monks Corner that I met later on.

A Buck Sergeant came out to give us instructions on what and how materials would be issued to us. I've learned that Buck Sergeants, E-5s, are the real tough asses in the Army. They're the ones who have just become NCOs and are trying hard to prove they deserve their new leadership roles.

"First, we are going to issue each of you a set of dog tags. These will provide your name and military serial number so that, if required, we can identify who you are."

"That's so dey can tell who you is if you's blown to pieces. Dey's a little notch in the ends of dose metal tags that can be put in you mouth, and crammed between you teeth so dat you mouth can be slammed shut and hardened after you is dead."

"Mills, that sounds stupid," Johnson said. "There has to be other ways to tell who you are when you've been killed even if you're not all together in one piece. Hell, if nothing else, they can look at your damn name printed on the front of your uniform shirt. Come on, let's go inside and see what they're giving away in this place." Johnson was trying to act tough, but he was as apprehensive as the rest of us.

The warehouse was huge—long rows of countertops with uniforms, boots, shoes, caps and hats piled everywhere. We were lined up and issued several sets of fatigues, which were to be worn during training every day. Footwear, when in fatigues, was black leather boots that had strong toes and back (Achilles) supports. The heel supports and toes were sewn into the boots separately. Also as part of the battle, or tactical, uniform were fatigue caps with rigid bills, floppy KP caps, helmet liners and "steel pots" (these were the outer, heavy helmets we'd all seen in war movies).

"These things are heavy," I said as I stood there with the hard helmet on my head, as we tried other stuff on and slowly moved down the line.

"Dumb butt, you're supposed to wear a *liner* under the steel helmet. You look like 'Beetle Bailey' or 'Sad Sack' in the cartoons with that thing hanging down and covering your eyes." McCoy was looking at me like I was something to be pitied, or was from another planet.

Honest attempts were made by the low-rank, inexperienced quartermaster personnel to provide clothing that was as close to the correct fit as possible. These fellows were certainly not experienced tailors. And, for those of us who were very tall, very short, way too thin, or heavy, it meant that major alterations were needed. For me, at five feet, seven inches and slightly over two hundred pounds, what seemed like a couple of feet of cloth had to be taken off my pants lengths, and my sleeves had to be cropped by several inches. It was difficult to imagine me being correctly fitted in this stuff, and looking "smart" in fatigues. Next, we were issued our "Class A," or dress-up, uniforms. The winter uniforms were wool and dark green with hard brim hats of the same color, and uniforms for the summer were khaki, worn with khaki caps. Low top, black leather lace-up shoes were worn with both. We were also provided with several sets of "swing easy"

underwear—not jockey shorts—two sets of long johns (I still have one set forty-eight years later, and my wife periodically tries to throw them out), socks, towels, washcloths and other everyday toiletry items. The term "swing easy" was somewhat of a misnomer, at least in the case of most of my buddies. We had little to swing easy with, which was undoubtedly made worse later by our training outdoors on cold days.

"You-all can throw most of that shit you brought from home away when we return to barracks," the Buck Sergeant said as he led us outside. Issued supplies that were not left to be altered were stuffed into large duffle bags, which were shouldered and carried back to the barracks.

The morning of our second day was drawing to a close. We spent that afternoon and most of the next day marching, marching, and marching. By the time we trainees left the Reception Station we were able to perform most formation maneuvers moderately well—without carrying firearms.

At ten o'clock each night lights were turned out in the barracks, except in the latrines, where men gathered to write letters back home. We looked forward to this because it enabled us to maintain contact with family and friends we had left behind. However, we were unable to *receive* mail while we were at the Reception Station.

The off duty time at night gave me a chance to write to my sweetheart, and think about why I was in the Army in the first place. The truth is, I had been a total failure in high school and during my freshman year in college. In fact, I had such a poor academic record and such tough-acting, wild behavior at Louisburg Junior College that the college Dean asked me not to return for the fall semester. My parents, Charles and Margaret Manooch, thought it best that I enter military service to gain some personal discipline, and think about just what I wanted to do with my life. Dad was a career officer in the Army, and he reasoned that after a stint on active service I would chose the military as my profession, or come home eager to straighten my life out by pursuing some other career.

I can tell you this. After just three days at Fort Jackson I was totally under somebody else's control, I was miserable and the time *crawled* by. There were—up to that point—two times in my life when I didn't know if I was going to live or die. One was when I was playing football at Broughton High School, and players had to practice twice a day in the mid-August heat without any break for water. We received no water during practice back then. At the conclusion of each August two-a-day practices we had to run wind sprints or repeated offensive plays at full speed. This was devastating in the heat.

The second time was before I played football and was when Howdy Manning and I were about thirteen years old and had a summer job inoculating chickens. That's right. Giving them shots. We were literally dropped off in the country at a big chicken coop located on a farm in Wake County. This was long before the time of the huge hog and poultry farms that are everywhere in eastern North Carolina today.

Howdy and I were directed to this large coop where it looked like a thousand white chickens were inside running around on a dirt floor with wood shavings on it. We were each given a small injector, for a better word, that looked like one of those double-pronged utensils you use to hold corn on the cob. Some type of medication was applied to these things. I don't remember how. The difficult part was running down a chicken, lifting her wing, and sticking the needles into the skin under her armpit (wingpit?). A complicating factor for us was that at first we had no way of determining which chickens had been injected and which ones had not. All the chickens were in one big area. Howdy was the first to figure this out. "Chuck, we need to do something to divide these chickens up." We decided that we needed some sort of a divider so we used a big piece of canvass to partition the coop into two big sections. I had hay fever back then, so the feathers and sawdust had me sneezing my head off. The day seemed to go on without end, and it was not too surprising when we didn't report back to work the next day.

Of course, the third time was active service at Fort Jackson, where there was no escape from harassment twenty-four hours a day, and there was no opportunity to mingle with the opposite sex. There were WACs (Women's Army Corps) on Base, but we considered them to be whores or lesbians, and viewed them with a different eye. Besides, we never got to go near them anyway.

The third morning of our stay at the Reception Station was *the* day. We were called into formation and individual orders for Basic Training were issued. Johnson, Lee, McCoy, Mills, and I were assigned to Company C, Tenth Battalion, Fifth Training Regiment. It was only a short distance away "as the crow flies," but immeasurably far away for a group of mostly teenage boys.

CHAPTER FOUR

The first few days in my newly assigned company were part of "zero week" because they didn't count for anything. The eight weeks of training could not begin until all the recruits had reported in. Like all of the Basic Training companies, Company C consisted of an orderly room where trainees never went unless ordered to do so, a supply room, a mess hall, and four barracks with each housing a platoon of recruits. There were two private rooms on the second floor of each barracks where the Drill Sergeants stayed. No one dared go in there either. The first day

I didn't have much to do except get settled in my barracks, and frankly it was boring and lonely. The few of us who had been together since arriving in Columbia were spread out—some in one barracks and some in others. Our little, familiar group was soon joined by several other fellows. One boy was extremely eager. He was a big, loud, dark-haired guy named Plank, from New York—or some other place "up North." To us he seemed as nutty as a fruitcake, and we called him "Gung Ho" or "Pvt. Plank." He loved the Army and all he seemed to want in life was go Airborne, and then join the Special Forces so he could go off to fight in some war. Another new fellow, and a complete contrast to Plank in personality, was named Gilcrest. He was also in the North Carolina National Guard like Lee, Johnson, and me, and was from the small town of Elkin. Later I realized that with his red hair and freckles Gilcrest resembled an older "Opie" from television's *The Andy Griffith Show*. One of my closest friends was a trainee named Barry Mahoney, a tall boy, about six-five, who had played basketball at Kent State. His name and mine were so close alphabetically that he and I were bunkmates. We were discussing bunk assignments. "You're the smallest, so why don't you take the upper," he said as we stood there looking at the simple metal-frame beds, each with a thin mattress rolled up on one end.

"The upper is fine with me," I said as I climbed up there to get the feel of things. That night, after chow, when several of us entered our barracks before lights were cut off we heard talking coming from the latrine so we went there to see what was going on. "What are you fellows doing," Gilcrest asked the four recruits that were squatting on the floor in their new green fatigues. I was afraid to ask, but it looked like they were gambling with cards and dice.

"*?Que pasa?,*" one said as he looked up, and they all laughed at what we didn't know.

"Where you guys from?"

"Puerto Rico. We're in the Army Reserves, and like you we've been sent here for six months active duty training. We've learned since coming to Fort Jackson that it's better to understand what you *want* to, and 'play the Spanish card' when it's helpful. That'll get you a long way around here." We all laughed together but soon learned that truer words had never been spoken. The soldiers' names were Castro, Lazada, DeJesus, and Quenonies, and our two groups started off on the right foot. Castro was small, slim, with sharp eyes and a strong, proud face. A week later he was appointed Squad Leader, and would wear a dark blue armband just above the elbow on his right arm to indicate his status as one of the trainee leaders in the barracks. Lazada was short, fit looking, and strong. He later proved to be very good competition during our physical training tests. DeJesus was also short, plump and seemed always happy. He smiled all the time and loved making others laugh. Quenonies was the tallest of the four and was quieter—even reserved. All four were very open and nice to us.

The next morning we were awakened at four thirty (or "0430" as we soon learned), had to run about a half-mile as part of PT (Physical Training), and then were marched to the chow hall. Outside was a wood and steel apparatus that resembled parallel bars on a kid's playground. It was "monkey bars"—consisting of a steel ladder at a height of about seven feet and held parallel to the ground by four large wooden posts. From that day on—every day—we had to swing from the rungs of the ladder and go from one end to the other before being admitted inside to eat. Most of the guys could do it fairly easily, but a couple of fellows, either very overweight or lacking strength in the upper body, dropped to the ground and were ordered to try it again. They were eventually fed, but had to work hard so that they could improve on the apparatus each day.

Early next morning, the Drill Sergeant formed the forty or so of us in ranks. We were wearing our new, dark green, unwashed fatigues. Most of the fellows had their names printed in black on a white cloth strip on the right side of their chests, and "U.S. Army" in gold on a black background on the left. Those of us who had our uniforms held over for major alterations had neither.

"All of you are being assigned to work details today. You'll remain with your detail throughout the day and return here when you're picked up. You'd better hope that your new companions arrive here *soon.* This time is *not counted* against your eight weeks of Basic, so you-all are merely 'treading water' until we have a full complement for this Company."

"I wish those other pussies would hurry up and get here so that we can get our guns and really start training."

"Plank, you're full of shit," McCoy said. "You *love* it here so much this will give you even more time to watch all of us be miserable."

When duties were called out, Barry Mahoney, three other guys and I were assigned to work in the supply room of another company that was well into its Basic Training schedule. One fellow carried work orders for our small group that contained each of our names and our service numbers. When we unloaded from the truck we were ordered to form ranks. We could tell immediately that this was not a very friendly place to be dropped off.

"Oh shit. Look who is going to be in charge of us today," I said as we stared at the big black Sergeant standing there thumbing through our papers. "See what Sergeant Johnson has on his shirt? You know what that set of wings with the parachute means don't you? Airborne. And, Airborne soldiers think they're tougher than the rest of us. He's going to give us hell today, and for as long as we are assigned to work under him. When an enlisted Airborne soldier meets an Airborne officer the enlisted trooper salutes and calls out 'Airborne, Sir!' and the officer returns the salute with 'All the Way.'"

"Okay, you bunch of dirty 'legs,' when I call your name step up here beside me." Sgt. Johnson was still looking through our orders, and had a smile on his face that looked an awful lot like a big, predatory cat. "Mahoney, Barry. You don't look too bad Barry M., but you'd better learn how to 'blouse' your boots." We all did. We learned how to tuck the bottom of our pants legs into our boots, and then fold back the loose fabric on either side of the calf so it would fit tightly in the boots giving the pants a smart appearance.

"Manook. Charles S. the third. What kind of damn name is that? It sounds like shit. Come up here and let me get a good look at you. I bet you think you're tough don't you Manook? Played football back home and liked getting in fights didn't you?" *Not as tough now as I was before coming here, I thought. There's no escape and they're on* your *ass twenty-fours hours a day! I didn't know what "tough" was up until that time.*

"Where the HELL is your goddamn name tag Manook?"

"They didn't get it sewed on yet Sergeant," I said as I looked down at the pitifully bare spot where the name ribbon ought to be displayed.

"I tell you what we're going to do. Pvt. Blanchard! Come out here right now!" A small pale face sheepishly peeked out from the doublewide, sliding wooden supply room door and blinked in the bright sunlight. The boy attached to it cautiously stepped outside and approached our formation.

"Blanchard, I want you to take one of those name ribbons you've got in there and print 'SHIT' on it, and then pin it on Manook's shirt. Manook, when I call out 'Pvt. Shit' I want you to come up front and center. When you report to me for work each morning you had better have on that nametag. You got that?"

"Yes, Sergeant," I said as I tried to look brave and stare at him square in the eyes. *I wonder what Dad would think of 'Manooch' being referred to as 'shit' since he was a Major in the Army? I know what he'd think. He'd probably love to see me being disciplined like this. That's what he'd think. "Margaret, our son's going to be a man when he gets finished at Fort Jackson."*

The five of us reported to Sgt. Johnson over the next two days and each morning the Sergeant seemed to take unrelenting pride in calling out "Pvt. Shit," and seeing me come forward like the "sacrificial lamb" to receive harassment—which was piled on me during the entire work detail. We were instructed to fold blankets, scrape and paint shelves, load and unload supplies, and clean up the supply room area and grounds.

"Police the area" became a standard order, which military personnel understand, and means pick up anything that litters the ground. I'd bet about ninety percent of what's picked up is cigarette butts. Our work detail was given periodic breaks, and offered the opportunity to "smoke 'em if you have 'em." I don't know why, but I didn't smoke. Most of the guys that I was stationed with did. My reluctance to smoke might have been deeply ingrained in my mind ever since Tommy Green gave me cigarettes down near Boone's Pond in Raleigh when I was about eleven years old. I got deathly sick after smoking most of the pack, and our neighbor Dr. Louis Kerman made a house call to find out what was wrong with me. Mother was not at all pleased when she found out the nature of my "illness" and took her frustration out on Tommy's mother, Barbara Green, who lived across the street. Mrs. Green, however, denied even the possibility her little Tommy could be the one who instigated the "smoke 'em if you have 'em" escapade at Boone's Pond.

The second morning, as we rode over to our duty company, we saw a group of soldiers working beside the road. "Why have those fellows got a large white 'P' whitewashed on the backs of their shirts?" one of us inquired as we turned to look at the miserable looking guys toiling by the roadside.

"They're prisoners. That's why," the driver said as he gave the work gang plenty of room as we drove by. "You think you've got it bad? You ought to see what those poor sons of bitches have to go through at the Base Stockade. Not one thread of kindness enters those locked gates. The lights never go off there, and the prisoners have to literally 'walk a line' that's painted on the floors." I learned later that most of the prisoners had been convicted of nonviolent crimes, such as being AWOL (Absent Without Leave) or failing to obey orders. More serious offenders were "sent off."

All of the troops assigned to Company C had reported by mid-week, and we could finally began the official eight-week countdown. We soon learned that when it comes to the Army, anything that's not official doesn't mean squat. By now our barracks were completely filled, and we were given precise instructions on how to make a bed the correct way and how to store personal gear in wall lockers and footlockers.

The barracks and an individual's personal gear were subject to unannounced inspections, and everything had to be in order to pass. If one trainee failed an inspection it could reflect poorly on the entire platoon, so peer pressure was tremendous. A guy who "fucked up" had better do it only once. Habitual slackers were severely criticized by their platoon mates. The beds had to be made so tight that a quarter could be bounced off it. The way the contents of a footlocker are arranged and displayed followed a longstanding protocol. Underpants were swing easies, not "jockeys" or "y-fronts" that some of the guys had worn when arriving. These and the high neck undershirts were all white. Each had to be rolled a certain way, and then placed on a white towel that was spread evenly in the top tray of the locker. The easiest way to roll the underwear was by using a beer can or soft drink can and roll the underwear over it. However, using a can to make a good display was against regulations, and if caught you were punished by being assigned to extra work details. Other displayed items such as a bar of soap, toothbrush, and razor all had to be arranged just right to pass an inspection. Field jackets,

uniforms, and boots were stored in the wall lockers, and they also had to be arranged in a precise manner to pass muster.

"When we getting our guns, Sergeant?" Mills asked Sgt. Riley one day while we were standing at ease in formation. We were eager to hear the answer because this was something we had wondered about since being assigned to our unit.

"When are you getting WHAT!? Johnson! Run over to the supply room and tell them to give you an M-1. And hurry the hell up." Johnson was back with a rifle within a few minutes. "Mills, get your butt up here NOW! Now look at me." The Sgt. held the M-1 up in his left hand. "From this time on you'd best remember this:

'This is your rifle and this is your gun (he grabbed his crotch). This is for shooting, and this is for fun.'" The entire formation burst out laughing.

"Okay, now I want all of you smartasses to extend your left hand up as if you're holding a rifle, and at the same time grab your 'Johnsons ' and repeat what I just said." That afternoon we were issued M-1 .30-06 rifles, which were stored locked up in racks on the ground floor of each barracks.

Following PT and breakfast the next morning we were marched to a large hangar-looking building to attend a class on the M-1, which actually proved to be very interesting. We learned that the rifle was one of the most significant weapons used by U.S. forces during the Second World War, and helped to achieve victory over the Germans and Italians in North Africa and Europe, and over the Japanese in the Pacific Theater. It was also used to great effect during the Korean Conflict. It is semi-automatic, gas-fed, and holds an eight-round clip, which is inserted by pushing it down into the open breech. When all eight rounds have been fired, the clip pops out with a "cling"- sound. The empty shells and clip are expelled to the right so the M-1 should be shot right-handed. If shot from a left-handed position, the spent brass and clip pop out and could hit the soldier on the cheek or neck.

The older NCOs spoke highly of the performance of the M-1 in combat. They said it was sturdy, easily assembled and reassembled under horrible conditions—could withstand water, mud, and sand—and was accurate up to three or four hundred yards. They said the M-1 was a much better weapon than the new

M-14, which was at the time being issued to Army personnel following graduation from Basic Training. The M-14 was reported to be less reliable than the M-1 under bad environmental conditions, and was less sturdy.

During morning and afternoon sessions the speakers stood beneath a six to seven-foot length model of the M-1, which was suspended from the ceiling. Following two days of classes and close instruction, we were able to disassemble and

re-assemble the M-1 blindfolded. It is a marvelous military weapon and ranks up there with the AK-47, 7-mm Mauser, and 1903 Springfield as being the "tops of the line" of light infantry weapons used since the turn of the last century.

Throughout Basic each soldier was required to study and learn the United States Military Code of Conduct. "The Code" is an ethical guide established during the President Eisenhower administration in 1955, and is a Department of Defense directive containing six articles. The articles outline how American military personnel should act in certain situations during times of war. In essence, U.S. personnel should evade capture, resist as a prisoner, and escape. Personnel should exhaust all reasonable means of resistance to the point where certain death is the only alternative, make every effort to escape and assist others, and never accept special favors from the enemy. Under the Geneva Convention, those captured should only provide name, rank, serial number, and date of birth to the enemy. The articles of The Code were displayed in the barracks and on the bulletin board outside. As we studied and memorized this official document it put everything we were doing in a very serious, even somber, perspective. We could go to war—and war is very serious business.

CHAPTER FIVE

"WAKE UP! WAKE UP!" I sat up, and still half asleep, stared around at nothing in particular, desperately attempting to focus my eyes and summon my brain into action. Slowly, what looked like a dirty face came into view as it peeked over the end of my bunk. One cheek bulged out like there was a golf ball in it.

"Remember, you're on detail today. It's three-thirty, so you'd better get up from there, get dressed, and go over to the mess hall."

Okay. *Now* I knew what was going on. The barracks fireman, a boy from Tennessee named Mumford, had the responsibility of seeing that recruits assigned to work details were up and ready to go by 0330, an hour before the rest of the troops got up. He'd long been up checking on the barracks coal-fired furnace. Those on detail had to wrap a towel around the foot of their bunks before going to sleep so the fireman would know who to rouse early the next morning. I remembered that I had placed the "towel indicator" at the foot of my bunk the night before.

Like the other firemen, Mumford had volunteered for his duty. Each barracks had one fireman who was in charge of tending to the furnace, and this required virtually twenty-four hour duty—shoveling coal and checking gauges. However, it allowed him to stay close to the barracks, and thus avoid some of the training outside in the cold. It was a hot and dirty job. The good thing about being a fireman was that for most of the time he was on his own, and not harassed to death like the rest of us. I guess, like most things, being a barracks fireman had its good and bad sides.

"Okay, thanks Mumford. It's not your damn fault that I have to get up before the chickens." He nodded and smiled at me with a wide grin partially exposing the big chew of tobacco held in his cheek. Small streaks of perspiration, highlighted by coal dust, extended from his forehead to his chin. He turned around and disappeared in the dark as quietly as he had arrived.

Three of us—Mahoney, McCoy, and I —had been assigned to work that day on a firing range where a company-size unit was to display the awesome firepower of both light and heavy infantry weapons in simulated combat situations. A truck picked us up right after our early breakfast, and ferried us to the range, which was located on the opposite side of the military reservation from our barracks. We huddled together in the back of the open truck to avoid as much of the cold air as possible. "Just think. Those guys are still in their bunks and we're already heading to work," Mahoney muttered. As we rode up to the entrance to the range a large sign proclaimed:

"I Am The Infantry.
I Am The Infantry—Queen of Battle!
For more than two centuries I have kept our Nation safe, purchasing freedom with my blood. To tyrants, I am the day of reckoning; to the oppressed, the hope for the future. Where the fighting is thick, there am I...I am the Infantry! FOLLOW ME!"

"Good Lord! That sure is something," said McCoy as he looked at Mahoney and me. We jumped down from the truck and each of us was assigned to specific work details on the range. Mahoney was carted off some place, and McCoy and I were taken way, way down range where we were greeted by an NCO and two other recruits who stood there looking woefully lost in their baggy, brand-new, never been washed, dark green fatigues—just waiting for someone to give them instructions.

"You guys come over here, grab one of those paddles there, and we'll get started." The Sergeant ordered each of us to stand behind a large open container that looked like an oil drum. Inside the drum was a thick pinkish-colored stuff that had the consistency of snow slush.

"This mess smells like alcohol," I said as I stirred it with my paddle.

"Naw. More like weak gasoline to me," replied McCoy.

"I ought to know. That's how we got our gas in Baltimore. We'd ride around and pick out a suitable parked car, and then we'd siphon out the gas." He was probably right about the smell. However, It felt good working there in the sun, slowly stirring and talking at the same time. No one bothered us, and we worked with the barrels—taking as much time as possible—until chow time.

The company in charge of the work detail delivered hot food to us right there at the range. The food was kept warm in large insulated containers that were white inside and olive drab on the outside. As usual, I couldn't complain about the chow. As we finished eating, an officer walked over to where we were sitting.

"You guys want to watch the demonstration this afternoon?" The Range Officer, a slim, ramrod straight Second Lieutenant was extending the totally unexpected invitation. *Hell yes we would*! That meant we could stay there, out of sight and mind, and pass away most of the afternoon and see all the firing. Two-and -a-half-ton trucks, called "deuce-and-a-halfs" began to arrive, carrying a company of soldiers who were going to be the audience and observe the firepower demonstration.

"They've already graduated from Basic and are in Advanced Infantry Training," said Mahoney. He was smart, seemed to know something about everything, and unlike me, he had successfully completed one year of college. That put him in a rather unique small group in the Army at that time. That was before the draft. The arriving troops were marched into the bleachers and then ordered to sit—all in unison.

"Welcome to the Fort Jackson Infantry Queen of Battle Live Fire Demonstration. What you are about to see is a thoroughly trained group of young men who will participate in an exercise to demonstrate the awesome fire power of a U.S. Army Infantry company." The Range Officer had obviously given this little introduction many, many times. He proudly pointed down range where we could see several targets, including one old discarded tank. From behind either side of the bleachers a couple of hundred soldiers streamed out and ran in front of the viewing stand where they took up positions at previously assigned locations.

"Look at that," McCoy said as we turned and stared at where he was pointing. Two dogs had followed the troops onto the range, and were standing looking at each other about twenty yards in front of us.

"Hey! Get the hell out of there!" The Lieutenant yelled as he waved his arms. He was obviously not prepared to welcome two dogs as part of his demonstration. The dogs moved closer to each other and began to snarl. The entire audience began to stand in anticipation of the fight that was sure to ensue. " Could be worse," the Range Officer said loudly through the sound system. "At least those mongrels are the same sex." Just to prove him wrong, one jumped the other from behind and started happily humping right there in front of the delighted assemblage.

"Good lord, look at that dude go. You can tell he's smiling from all the way up here."

"Shut up McCoy. Give due respect to two animals caught up in a passionate moment of making love," Mahoney said with a big grin.

"Making love my ass. They're flat out screwing. You can call it what you want to." McCoy shot back. The dogs couldn't maintain their frantic pace for long, and after only a minute or two they separated and went off in opposite directions. Both were panting with their tongues hanging down, and they briefly looked back at each other as they strolled off. Thus, a quick romantic encounter was soon forgotten.

The audience was still buzzing over the lewd display when a huge explosion down range ripped through the clear, cool afternoon air. The shock of the loud boom and waves of warmth rippled through the stands. All eyes focused on the row of barrels that were spaced at intervals about three hundred yards in front of us. One of the barrels had been flung into the air and had returned to earth surrounded by a huge reddish-orange fireball.

"Jesus. *That barrel* was one of those we were working with," I said as I looked at McCoy who was standing there with his mouth agape.

"You mean that you guys were off all day playing with *napalm?*" Mahoney questioned with a surprised look on his face.

"Napalm? We were just doing what we were told to do, and that was fill those damn barrels and stir that smelly stuff up," I responded.

"Well, you can see what that 'stuff' can do don't you? Just think what could've happened to you two if one of those barrels had gone off while you were standing there. You'd look like little black piles of burnt pork, that's what. No disrespect meant McCoy."

"None taken," said McCoy. He and I both knew that Barry Mahoney was too nice of a guy to have meant anything racial.

We didn't have long to discuss the horrific effects of napalm because at that time the range erupted with the sounds of rifle and machine gun fire. A jeep carrying a 106 recoilless rifle pulled up and fired a .50 caliber tracer down-range to confirm the target. The 106 then blasted the disabled tank down range. The audience sat in awed silence as M-14 rifles, M-79 grenade launchers, M-60 machine guns, mortars, a flamethrower, and bazooka were all brought into action. When we left the range that afternoon we were greatly impressed by the weapons demonstration, the cohesion and skill of the troops that took part, and at the same time we wondered if we would ever reach that level of capability.

That same week the entire company was marched to rifle firing ranges located about five miles from the barracks. We were excited because it was our first opportunity to fire the M-1, which we had studied, cleaned, disassembled and reassembled over and over. Each recruit was required to fire the rifle in prone, sitting, and standing positions. The shooting stations, or positions, were marked with white wooden stakes that were driven into the

ground. I already knew from my previous experience at Camp Butner, NC that the prone position offered far more support and steadiness than the other positions when you didn't have something to rest the rifle on. Therefore, firing lying down was usually more accurate than standing or sitting.

"Ready on the right! Ready on the left! Ready on the firing line!" were the three commands issued by the Range Officer from the observation tower, which stood behind the firing posts. His next command was normally "Commence firing!" However, before he could get the words out there was a loud *"KA-RACK"* as a recruit fired prematurely. "Get that son of a bitch off that line," the officer screamed at the NCOs who were walking up and down the firing positions instructing the trainees. The poor fellow was jerked up and marched off for punishment. None of us looked at him; we just stared straight ahead until we were given orders to fire.

Bull's-eye targets were placed at one hundred, two hundred, and three hundred yard distances. These targets weren't intended to provide a very realistic combat situation, but were used to evaluate, and hopefully improve, the marksmanship of the trainees. Some of the guys had never fired any type of weapon before. We fired at the stationary targets during one training session only. Afterwards, most of our Basic Training firing exercises took place at the newly installed TRAINFIRE 1 Range. Different aspects of this range introduced realism to the training situation. Recruits had to quickly identify and estimate the distance of pop-up targets that resembled enemy riflemen hidden in wooded areas. The trainee had to be ready to fire at any time, and do so quickly from the most convenient position available to him. In these instances accuracy was sometimes reduced in order to provide the most firepower possible. In combat it is very important to provide enough firepower to keep the enemy pinned down and unable to adequately return fire. If you hesitate to return fire, a lot of bad things can happen.

In modern warfare the amount of projectiles flying through the air is what counts. Heavy firepower tends to reduce the effectiveness of enemy infantry by causing them to "hunker down," not return fire, and importantly—not move. It is interesting to review military history and discover how many bullets were fired to produce a single casualty or fatality. The following numbers are not accurate, but do adequately reflect the idea of increased random firepower. In the American Revolution one hundred bullets were fired from flintlocks to cause one hit. In the War Between the States, one thousand projectiles from cap and ball muskets were spent to produce one hit. In the Second World War the numbers were ten thousand for one; and in Viet Nam one hundred thousand to one casualty or fatality. Of course during that time—from the late 1700s through the early 1970s—the methods of infantry engagements have changed drastically from marching abreast towards the enemy to ambushing, evading, and attacking again. This evolution in infantry against infantry tactics has made hitting the target far more difficult.

Days were getting shorter and nights longer as winter approached, and temperatures were decreasing—rapidly indicating that we were in for a colder than average winter. This was not good, since we spent ninety percent of our daily time outdoors during Basic Training. The mood in our barracks reflected the dreariness outside, and one morning during PT we felt the light touch of the first snowflakes. It was still pitch black and the smell of coal smoke filled the cold air.

CHAPTER SIX

Thanksgiving was just around the corner and the fellows in Company C, Tenth Battalion, Fifth Training Regiment were getting excited about the big day. However, the excitement was greatly tempered by the fact it was to be an *off duty time only*—no one would receive a "leave" to travel home and spend the holiday with families and friends.

This marked the first time in my nineteen years that I was not going over to my grandparents' house to have the traditional Thanksgiving dinner. Estie and her husband, J.W. Callahan, were my mother's mother and daddy, and they lived on Vance Street in Raleigh just a couple of blocks from Five Points. When I was growing up we all ate together. Mother, Dad, my younger brother Chris, my grandparents Estie, Granddaddy, Loulie, Big Daddy, and my aunts and uncles, Maude, Stewart, Annie Laurie, and Uncle Truman gathered around the same festive table. The entire family lived within a mile or so of each other in the Hayes Barton and Budleigh sections of Raleigh. My brother and I were the only children, so you can imagine that we were spoiled. However, I was eleven years older than my brother, and therefore had a much longer time to get that way.

My sweetheart—when I was a teenager and for many years after—was Genevieve Jeffreys, a Raleigh girl three years younger than me. We had been dating for several years. She was well liked by my family, as I was by hers, and she joined us on many family occasions. On Thanksgiving we typically ate at noon with my grandparents, then she and I went to her parents' home to eat with her family later that day. Mr. and Mrs. Jeffreys lived just outside of Raleigh off Poole Road and owned about twelve acres of land, which included a small fishing pond located down the hill from their house. Mrs. Jeffreys—like my grandmothers—was an excellent cook, and at Thanksgiving she served the traditional southern dinner of turkey, dressing, gravy, collards, cranberries, snap beans, mashed potatoes, seven-layer jelly cake, coconut cake, and sweet potato pie. However, her much-appreciated specialty was oyster dressing, and I looked forward to this treat each year.

Oyster dressing is a favorite of mine. However, I haven't been able to duplicate the oyster dressing that Mrs. Jeffreys placed on the table. The Lord knows I've tried hard enough over the past forty-eight years. My oyster dressing never has the right consistency—it's always too dry, and the oysters have been cooked away so that the distinctive oyster smell is the only thing that remains in the dressing. My children carefully explore the stuff on their plates, look pitifully at me, and ask: "Where're the oysters, Daddy?" Heck, I don't know what to tell them. They're just gone. That's all. Disappeared into thin air.

Although we would not be together over Thanksgiving that year while I was at Fort Jackson, Genevieve and I were good about communicating by mail. I have a stack of letters today that I wrote to her while I was on active duty. The letters were postmarked from Columbia, SC, and each was mailed with a four-cent stamp. Genevieve's Raleigh address was "Route 2, Raleigh, NC" and required no zip code.

Daily mail call was a big deal in the military. It allowed us to receive letters, and thus maintain contact with the outside world. Letter reading and writing were usually done at night in the platoon barracks latrine after lights had been cut off everywhere else. Some of the fellows in my unit couldn't read or write, so it was a frequent practice for one of us to write down what a guy said and then mail the letter off, or to read a letter that had been received. This was offered as a service of friendship, but it got out of hand sometimes. Letter contents could be composed or read very differently from what was intended. Romances could be falsely nourished or devastated by a "friend" doing a *good* deed. Since then, I have come to realize that this creative trickery was truly "low life" behavior on our part.

On Thanksgiving all of the troops in the U.S. Army, at

least in garrison, around the world are served a very good dinner with the traditional baked turkey and all of the trimmings. We were even given second servings in our company mess hall, and could take our time and eat as much as we wanted. The following Sunday morning we were marched off to church. There were Catholic, Protestant, and Jewish chapels on Base. Most of the guys in my barracks called themselves Protestants. I never met anyone in the Army that referred to himself as an atheist. I don't know whether this interest in religion is because going to chapel allows a recruit to take it easy for a few more hours, or because when you think there is a good chance of being killed—like in the Army—you have to believe in something, and being an atheist in my opinion falls dreadfully short of that. It doesn't seem at all wise to me to eliminate at least a chance of salvation right from the start.

Just prior to going to bed we took showers, wrote letters, shined our boots, or messed around. Later, some activities, like shinning brass, could be done in the latrine where the lights were left on all night. This daily routine continued throughout our training.

One evening things were getting lively down at the other end of our bay. "What's happnin? Whassup?" The two black guys, McCoy and Ferguson Moody, on our bay would start jiving and cutting up, and it wasn't long before most of the white fellows were gathered around one of their bunks. These guys were always upbeat, and had us laughing in a matter of minutes. It was the first time that most of us from the South had been that close to Negroes, and certainly the first time on an equal basis. One of the most amazing things to me is how a couple of black guys could bring us all together by laughing and having a good time. It was magnetic.

"Know what we are going to do tomorrow?" Ferguson Moody said as he began to shuffle a deck of playing cards.

"No. What?" I replied as I squeezed in to sit on one edge of the bunk. I never played much poker on active duty, which made me different from most, but when something was going on I was generally right in the middle of it.

"Well we're getting up at 0430, as usual. Then we have our morning PT run around Tank Hill, as usual, and then we go to eat breakfast over in Sgt. Bayliss' mess hall. Again, as usual."

"That don't sound no different than what we do every damn day," Mills said with a smirk.

"You didn't let me finish, smartass. '*Then.*' The proper word is 't-h-e- n.' We are going to a class on first aid. That is if anyone here cares enough about his buddies to help him when he's been injured—like getting shot."

"I guess we *all* would help each other, wouldn't we?" said Mahoney. And each of us agreed that we would. The mood had turned from jive to serious in a hurry.

The bleachers at the first aid training area were made of weathered wood and steel, and you can bet that many a recruit's rear end had sat on them. Our platoon sat there in the cold, all huddled together with field jackets and gloves on. The gloves were unlike any I had seen. They were black leather with olive drab colored wool inserts. The bare branches of stunted hardwood trees around us waved back and forth in the wind, and made the stark landscape appear even bleaker. The sun was off hiding someplace and was nowhere to be found.

In front of us was a rectangular area of about fifty feet

by thirty feet covered by sawdust and bordered by two logs laid lengthways on top of each other, forming a boundary. Every outside class was held in a similar "theater." The Sergeant in charge of instruction was explaining what items were contained in a GI's first aid kit—mainly bandages of different types and sizes and various ointments and salves. The one medication that was obviously missing was morphine, but we were given a brief instruction on its use. Over the next hour volunteers were asked to come forward so that proper application of bandages could be demonstrated.

We learned that one of the most severe wounds occurring in combat is a "through and through, sucking chest wound." This happens when a bullet goes completely through the chest and leaves holes on the front and back of the torso. The "sucking" term applies to the fact that oftentimes a lung has been punctured.

Bullets used in the military back then were very different than those many of us used in hunting. When hunting deer with a rifle, I used a soft tip bullet that expanded upon impact, and therefore caused immediate damage to tissue and bone. Military bullets used in 1962 were steel-jacketed and would penetrate much deeper. They could even hit an enemy combatant after going through intervening objects, such as a person, large tree limbs, and boards. The negative side to this, however, is that there is a higher probability of casualties from friendly fire ricochets. Today, the military uses frangible bullets that cause extensive damage to the target, and are not as subject to ricocheting as are steel-jacketed bullets.

In combat, the chest wound is frequently caused by rifle fire, and is treated by applying a large square bandage pad to the victim's chest and another to the back—both held in place by bandaging wrapped around the upper body.

"How in the hell else would you patch up a guy that had been hit like that?" whispered Mahoney. "Do you mean we're sitting out here in the cold to learn something so obvious?" The instructor heard Mahoney's comments and was not at all pleased.

"Okay, know-it-all. Get your damn ass up and come out here front and center! All of you pay attention to what 'Dr.' Mahoney is about to demonstrate." Mahoney sheepishly raised himself and worked his way down to the front.

"*Dr*. Mahoney and I are going to introduce you to the next type of wound, which can be very serious and is caused by a phosphorous burn. Do any of you know anything about white phosphorous? Have you seen in a war movie a large explosion with hundreds of white streaks coming from it? If you have, then you have seen white phosphorous, which is usually delivered by mortar or artillery rounds. If any of that shit hits you it starts to burn tissue upon contact, and doesn't stop burning until it has gone completely through—whether it's your arm, hand or chest. The only thing that will stop this progressive burning is to keep air from getting to it—like by applying water, or some other liquid. If you've got a lot of water nearby then you're in luck and can start pouring it on the injured area. Isn't that right, *doctor*? If there's not much liquid available then you can treat it by covering the burn with an airtight mudpack. Mahoney, Barry here is going to show us how to make a mud cake without the use of water," the Sergeant said as he handed Mahoney a small metal bucket. He continued: "Mahoney, let's say you have a burn on your lower right arm."

"What am I going to do with this? It hasn't got anything in it." Mahoney was holding the small bucket and was looking at it suspiciously.

"Why, you're going to piss in it. That's what," the Sergeant said as we all broke out laughing. Mahoney made a deep, prolonged sigh, unbuttoned his pants, pulled out his penis, and strained to get something started.

"We're not leaving here until there's pee in that bucket."

"Come on, Mahoney," we pleaded. "You can do it." Mahoney's face got red and we couldn't tell if it was from his straining or embarrassment, but after a moment or two we heard the distinct tinkling of urine on metal.

"I ain't putting that pee on me," McCoy said as he made a face and recoiled with disgust. You can bet that the entire assemblage watched intensely as the Sergeant instructed Mahoney on how to scrape up dirt and mix it with the pee to form a wet mud cake, which was to be applied to the injured area. The ground was almost frozen so Mahoney had to dig hard until he had enough dirt to mix with the urine.

"If I was a damn medic, I'll be damn if I'd touch that mess on his arm until it had been washed several times," Lee said.

"That's the underlying problem, dumb butt. There's no water available," Johnson whispered.

"That'll keep the phosphorous from burning deeper and will hold things until you can be evacuated and treated by a medic or doctor," the Sergeant instructed as he held Mahoney's arm up so that we could all see.

"That phosphorous is some mean shit," Lazada said to those sitting close to him.

"Yeah, and when you think about that napalm and this stuff together you realize just how mean people can be," I said as I envisioned huge balls of fire and explosions with hot white streamers pouring down on a bunch of nineteen-year-old boys. Teenagers, little more than children who happened to be in uniforms, off somewhere fighting while folks back home discussed the theories and politics of yet another war.

As I've gotten older my attitude towards war has changed—and I guess it's only natural that I've come to the realization that those who die represent three very different groups of young men. Most are those who go with the flow, doing what they are told to do, and go about their business until their times in service are up. At one extreme are those who eagerly follow orders, go bravely ahead, and willingly lay down their lives for their comrades in arms. And at the other extreme are those who have been drafted or join up and have no business whatsoever being in the military in the first place. They are oftentimes the poorly educated—even mentally slow. I served with some of them at Fort Jackson. The movie *Forrest Gump* has as its main character a fellow who is simple, sweet and kind. And he's caught up the war in Viet Nam. He doesn't know what he's gotten into. When he's wounded he thinks he has been stung by a swarm of bees. Yet, he is dedicated to his friends and returns to the battle time and again to carry wounded comrades to safety. There are, and have forever been, young men like Forrest Gump serving their countries during wartime. Killing or injuring them is like hurting an innocent child. In essence, the only thing they have to give is themselves. Sadly, it happens much too often that this is their fate.

CHAPTER SEVEN

Soon after Thanksgiving I decided to explore the possibility of taking advantage of the different training opportunities available to recruits—as advertised—in the U.S. Army. Granted, the reservists and National Guard fellows could not be offered much since we'd be on active duty for only six months. Johnson, the boy from Archer Lodge, was sharing some of his thoughts along that same line. "You know, being a military driver would be a good thing. They have it easy. Drivers get up earlier than the rest of us, but all they do all day is drive somebody or something around. And, they don't have to sit outside in the cold like we do for most of the time now—they just park off to one side and wait there with the windows rolled up and the heat on until it's time to drive some place else. I don't know about you-all but that sounds right damn good to me."

I was listening to Johnson's reasoning and had to butt in on the conversation. "I'll tell you what. I'll do it if someone will volunteer with me," I said as I looked outside the barracks window.

It was cold and very windy. Small puffs of fine sand were swirling around on the Company street and added to the bleak landscape. We were waiting inside—something we very seldom had the opportunity to do unless we were asleep—before being marched off to the mess hall for chow, where the main course of the day was rumored to be fried beef liver.

Mother never attempted to serve us liver like some mothers in my neighborhood did when we were growing up. She thought that the stuff looked awful, felt awful—and considering where it came from—*was awful* and had no place at her table. It didn't really make any difference, because Mother wasn't a very good cook anyway. And, she never claimed that she was.

The Army served beef liver on a regular basis. At first I turned my nose up at it, but before long I honestly enjoyed eating liver cooked in onion gravy and served with mashed potatoes. It tasted good, was warm, and it filled you up. Those things are very important when you're in the Army. We had liver and onions that day as we continued our brief discussion about volunteering for driver training. Everything had to be brief in the mess hall. Once you got your food you were given little time to casually sit at the table and enjoy it before a Sergeant blew his whistle signaling the order to form in ranks outside.

"When we fall out in the morning, and after PT, the Sergeant will ask for volunteers for various OJT (on the job training) categories. That's when I'm going to stick my hand up," I said as I looked over at Johnson, hoping he would volunteer with me. He indicated that he would.

At exactly 0430 the following morning our four platoons stood in formed ranks in front of our barracks shivering in our t-shirts, combat boots, and fatigue pants. Trainees who were sick responded to "sick call," and were sent to the Orderly Room to be rather unprofessionally examined, and questioned to decide whether *real* medical attention was required. Being allowed to visit the base hospital was a treat, because the recruit could avoid at least a part of the day's training—and far more importantly—women were there. Good looking women who smelled good, and filled out their little white uniforms that twitched from behind when they walked.

Running a fever was a sure way of being temporarily dismissed from duty and more often than not, resulted in a trip to the hospital. Many of the guys in my platoon could not read a thermometer. I could, and was experienced at taking one of the older varieties and shaking it up so that it gave an artificially high reading. I usually tried to achieve a body "temperature" of about 101 degrees or so. Toby Lee had one stuck in his mouth that morning, but no more had a fever that I did. I read it and it was normal, about 98.6.

"Here, give the damn thing to me," he said as he grabbed it and started rapidly shaking the thermometer up and down. "Now, what does it say?"

"It says you're a dumb ass because you've got a temperature that's 107 degrees. Hell, that's higher than a duck's. Nobody would believe that your fever could be that high and you still standing here in a non-convulsive state," I replied regarding him with disgust. Neither of us went on sick call that morning.

"Okay, you guys. Pay attention. Before we go on our little run this morning I want to know if any of you would like to become drivers. Show your hands."

Johnson and I responded to the Sergeant's request and were ordered to report to the Company's Orderly Room following breakfast. From there we were driven to the Fort Jackson Motor Pool, where we spent part of the day in the classroom and the rest actually test driving different types of trucks. The motor pool was huge, and all of the military vehicles were lined up in perfect rows. Those we were to drive could be identified as either tactical or non-tactical. Tactical ones look like something you'd drive right on the battlefield; non-tactical are those you'd drive around town or the Base. Johnson and I had the opportunity to train using the quarter-ton jeep, the three-quarter- ton 4X4, which looked like a cross between a jeep and a truck; a one-and-a half-ton truck; and a two-and-a-half-ton truck also called the "deuce-and-a-half." All but the one-and-a-half-ton trucks were tactical vehicles. I preferred the deuce-and-a-half, which came with either a straight or automatic transmission. While test-driving with an instructor, I realized just how far I had come in four years since I took driver's education at Broughton High School one summer when I was fifteen. Coach Olin Broadway was our teacher and he made us drive a straight transmission car—something I was not prepared for and was actually afraid of doing. Here I was in the U.S. Army volunteering to drive all types of vehicles, and most had straight shifts.

Prospective drivers at Jackson were required to take a very short defensive driving course. This was one of the most practical courses I have ever taken, and when operating my personal car today depend on some of the skills learned during that course taken more than four decades ago.

New drivers soon learned that driving assignments could be very different. Perhaps the easiest duty was picking up an officer in a jeep and driving him wherever he had to go that day. Jeeps were generally open, so it was cold and required wearing a field jacket at all times. Drivers also had to know when to stay in the jeep at a stop, and when to get out and accompany the officer. Most driving details, however, required driving one of the trucks—as part of an ammunition convoy, conveying troops from one firing range to another, or delivering chow to troops in the field. I was at first apprehensive about driving around in a truck loaded with live rifle ammunition and mortar shells, and which had very large red signs on the bumpers with "EXPLOSIVES" painted in yellow on them. I used to wonder what would happen if the truck blew up. In other words, how far I would be flung, and just how much of me would be left after all the fire and smoke cleared.

Since military infantry training was the main purpose of Basic, recruits could not be taken away from daily training schedules too often to merely transport people or things from one place to another or, for other reasons, to cause them to be absent from their training duty stations. A soldier who missed too much training was recycled, and had to start Basic all over. Drivers were very seldom detailed more than once per week, and on those days they usually started the day at about three-thirty in the morning and ended around five in the afternoon.

One of the benefits of driving was that you could see first hand the fundamental training operations going on all over the Base, including the infiltration course, escape and evasion, night river crossing, bayonet drill course, live fire squad tactics, and the gas contact chamber. Before I left Fort Jackson I was exposed to all of these areas of infantry combat training, as well as some others.

We had just completed our third week in Basic when I decided to try out for the Fifth Regiment boxing team. My father boxed on the N.C. State College boxing team in the later 1930s when it was an approved college-level sport. I wanted to make it a part of my Army experience.

The gym was not far behind our barracks and the best I can recall it was in an old brick building with all the windows built in high near the roof. McCoy, Mills and I were the only three guys from our Company who reported for the first practice. We foolishly thought that if we were on the team we could be excused from some of the Basic training schedule. Much to our disappointment we learned that practice was held at the end of the day—after regular training. We didn't miss anything, but what we *did do* was get the living hell beat out of us.

"Manooch, you don't know 'shit from shinola.' Got us over here where we have to work twice as hard as we did before. I don't know about you." I could tell that McCoy was only half serious, because he really wanted to box.

"Well at least you've learned how to pronounce my name right. That's at least something," I said to his back as he stepped inside the gym door where the outside smell of coal smoke was replaced with the pungent odors of sweaty gym clothes, leather and canvas.

The regular fighters on the team were NCOs who had extensive training and had boxed for years. They smiled when they saw us come in. They were actually only smiling at me. I was the lone white guy in the gymnasium, if you count Mills as being half and half. The black guys and Puerto Ricans were from up North, and during the winter they spent most of their spare time in gymnasiums where they boxed and lifted weights. Mills was from the South—South Carolina—but he was just naturally mean as the devil, and that made up for his lack of formal boxing training. He'd certainly been in enough street fights. Several "boxing Sergeants" stood there taking a temporary break from jumping rope, passing around large leather medicine balls, and working out on punching bags.

"Come over here. Let's get your names and weights and then we'll see what you can do," one of the Sergeants said as he grabbed a clipboard.

The three of us loosened up before being observed in the ring. When I stepped into the ring that first time I was not only lacking in experience, but I was also lacking in height for my weight class, which was close to two hundred pounds. The fellow I fought was about one-ninety and he was at least six or eight inches taller than me. I couldn't get near him. He jabbed me silly, and when he wanted to he'd throw a punch that meant business. I have never been so stunned in my life, and if it had not been for the protective headgear we had to wear I'd probably still be in a coma. There were three two-minute rounds in an amateur fight. Each seemed to last forever, and the endurance required to finish each round was far more taxing than my previous experience playing high school football. Unlike football, in boxing the fighters are being hit in the face and head almost constantly. If your opponent is good, which mine always was, there was no way to escape the constant pounding.

McCoy and Mills fared much better, mainly because they were less heavy and therefore fought in lighter weight classes, which they took advantage of because of their heights. They were at least competitive and represented the Regiment well. I remember sitting blurry-eyed on a stool in my corner, spitting blood and water into a large metal funnel pained red, white, and blue, that was attached to a hose that ran off someplace. Several hundred recruits had marched over to the Base Gymnasium to watch the fights. This spectacle was a time off for them so they were all fired up and hollering. When all was said and done, I had two fights. I lost one, and the other was a draw. That was the highlight of my very brief boxing experience—and I never put on gloves again.

CHAPTER EIGHT

The long-dreaded Physical Training test was just around the corner, and while some of the guys were almost gleefully looking forward to it, I sure as hell wasn't. I harbored an inner fear that only a few of the fellows knew about. Most of the test was fine with me, but I was actually *afraid* of the one-mile run. I didn't like to fail anything that had to do with strength or conditioning. I could handle the other events better than most. I was strong, fast, and agile. What I wasn't good at was running a long way, and for me, anything over a quarter of a mile fell into that category. Hell, I had enough trouble running around a couple of blocks each morning during PT, and sometimes several of us schemed up ways to avoid the morning run. We'd volunteer for anything to get out of it, and on extreme occasions we'd slip out of the platoon formation as our buddies ran down the street, hide, and then rejoin them as they jogged up to the barracks. This was very risky because getting caught meant embarrassment and punishment. However, it was still dark during our runs so we were able to do this without being detected. I was daydreaming about running as we waited for the trucks to arrive.

"Manooch. I don't know why you're dreading the PT test. Hell, you're strong as an ox and you look like an athlete, although a short one I have to admit."

I glanced over at Lazada, that little snot, as our Company was being loaded into one-and-a-half-ton trucks. "Well, I may be short, but I'll score better than your little brown ass." This was one time we were not required to march because our Company cadre wanted us to perform well on the test and reflect well on them, and therefore we were given as much rest as possible before the testing started.

I don't remember all of the events each recruit had to enter. I do recall the "hand grenade throw," the "shuttle run," the "low crawl," and the "mile run." The hand grenade throw was just what it was called, and required accurately throwing a dud hand grenade. Each soldier was awarded points based on how close he came to the target. At this stage of our training none of us had even held a grenade. It was heavier than expected, and you sure couldn't throw one like a baseball as sometimes depicted in war movies.

In the shuttle run a recruit had to run—sprint if possible—

through a maze of waist high barriers, which were positioned in a serpentine arrangement. The shorter and faster fellows generally did better on this course because maneuverability was a critical element.

The low crawl resulted in my best performance, and several of the Sergeants told me later that I had set a new Base record. In this event five or six soldiers were spaced side by side as they would be at the start of a hundred yard dash. The major difference was that each soldier was required to crawl about twenty yards on the wet sand while lying on his stomach. I believe my time was about twenty seconds. However, I didn't get much of a chance to gloat about my victory.

"Second platoon, line up with me." Sergeant Jenkins was positioned at the starting line of a cinder-covered running track. Large puddles of water were left standing where a cold rain had drenched Fort Jackson the previous night, and the sunlight was reflecting off of them.

"This is where we separate the men from the boys," Lazada said as he proudly looked at our platoon, and it would be hard to miss the confidence oozing out of him.

"Sergeant, don't we get running shoes, or tennis shoes, or something like that?" one of the guys asked. I don't believe that I've ever seen someone as white as he was—standing there with no shirt on. The Sergeant looked like he couldn't believe what had just been asked.

"Dixon. Put your goddamn t-shirt back on, you dumb ass. And as far as your question is concerned, you're a damn soldier. How many United States Army soldiers have you seen in movies fighting a war and *running around in little tennis shoes?* That's something you'd expect to find on "chinks" and "wogs.""

You-all are going to run like you do in PT each morning—with combat boots on. That's how you'll fight and so that's how you will train. Come on now you guys, strip down to your t-shirts and let's get started."

This was the moment I had been dreading for days. The mile run!—a distance that covered four laps around the quarter-mile track. How in the world was I going to run that far? I just wasn't built for it. I didn't know it then, but the trainees that looked like pond scoggins—tall and skinny—were the ones best equipped for running the longer distances, and some of them looked like they'd done this competitively before. They leaned slightly forward and held their arms in a running position as they approached the starting line.

"Get on your mark. Get set. Go!" Sergeant Jenkins commanded, and we were off like a pack of wild animals splashing out of the starting gate. I had sprinted while in high school, and I didn't like being behind, so I dashed into the lead. This position was relinquished after a short distance, and realizing that I had literally "shot my wad" I abruptly slowed to a jog. The flock of gangly wading birds ran past me kicking up cinders and splashing water—and none of them even glanced my way.

I guess I covered about three-quarters of the first lap before I had decelerated to a walk. That was a very bad mistake.

"Manooch, what in the hell do you think you're doing?" One of the other Sergeants had run across the infield to catch up with me. "This is a damn *run* not a *walk*!" I noticed then that I seemed to be the only contestant that was walking.

By this time I was breathing so hard it was difficult to

understand what I answered. "I can't Sergeant. I think I'd drop dead if I tried to run."

"I'll see that you drop dead if you *don't*. So pick it up." He stayed with me step for step until I had completed the run—although our pace was pitifully slow.

To achieve a score of one hundred percent, a trainee had to run the mile in about six and a half minutes. It took me almost ten to walk and jog the entire course. My score was so low that it pulled my overall PT test results way down. Members of our platoon gathered behind one of the trucks that would ferry us back to the barracks. Most of the fellows were breathing hard and shivering, and at the same time actually sweating, although the temperature was just above freezing. "What do you think you made on the test?" Mahoney managed to say as he gasped for air. He was leaning over with his hands on his knees, and wasn't speaking to anyone in particular.

"I think I done pretty good. Dem tests weren't that hard, but I sure throwed that hand grenade to the wrong place." Mills was like McCoy, in that both of them never hesitated to offer an opinion no matter what subject was being discussed.

"Well Mills, you might have done okay, but being as tall as I am I didn't fare too good on that shuttle run or the damn low crawl. It seemed to take forever to get things moving." Mahoney was right. He'd probably done well on the mile run because of his college basketball experience, but the two events that required quickness as well as speed were his undoing.

"How'd you do, Manooch?" Lazada said this with a smirk. He and I had a continuing argument about which one of us would score best. He'd seen me at times walking the track so he figured that event's score would be low enough to enable him to move ahead of me in the overall results—and he was right. He placed first in our platoon. I was third. And as for Company C. Lazada was second and I was about sixth. Not too bad considering my mile "run" had scored only forty percent. Lazada was not a very shy fellow, and he took every opportunity to tell anyone who'd listen what a fine physical specimen he was. But, you know what they say—"what goes around comes around."

One of the duties we had to perform at least once every day was to thoroughly clean our M-1 rifles. A soldier's life could literally depend on the condition of his rifle—a fact stressed to us constantly by our cadre. Each of us had an assigned M-1. We had to know its serial number, and have the rifle ready for inspection at a moment's notice. Some of the ugliest scenes that I witnessed while standing in formation were the results of improperly cleaned weapons. Naturally, we took this very seriously, and spent hours of treasured spare time cleaning and oiling our rifles.

The night after our PT test, trainees in our bay were busy with oil, rags, and cleaning rods. The oil came in large five-gallon cans that were—naturally—olive drab in color, and each can had a screw-on cap that was about two inches in diameter. The cleaning rod was in the butt end of the rifle stock. Our favorite athlete was sitting on the floor like a monkey and was trying to open a new can. "Damn, just when you think you've got this thing open there's something else to deal with." Lazada was referring to the thin piece of tin covering the opening that remained and blocked the opening even when the cap was first removed.

"Well, get a damn screwdriver and punch a hole in it," McCoy said as he walked over to where we were gathered. Unfortunately, that's exactly what he did. But rather than use the screwdriver to punch the hole and then pull the tin tab out, Lazada put his finger in the hole to pry the tin free. That was a hell of a mistake. His finger got caught tight, and when he tried to extract it the sharp edges cut into his skin. Just as we were contemplating what to do to help Lazada we heard someone coming up the stairs.

"Lights out!" Sergeant Riley stood at the head of the steps with his hand on the light switch. Then he saw Lazada. "What in the name of God are you doing Lazada? Why have you got your finger in that can of oil? Come over here and let me see."

"I can't Sergeant, the can's too heavy," Lazada moaned with a pathetic look on his face as he knelt on the floor.

"Well leave the damn thing on there like it is until morning, and then we'll see what can be done. All of you get into your bunks! I'm turning the lights out now!" His instructions to us reminded me of a big slumber party—in a horror movie.

A five-gallon can weighs roughly forty pounds at eight pounds per gallon. That's a pretty heavy weight that could be left hanging unsupported from a finger all night, particularly if you had to sleep on the upper bunk like Lazada did. Man, if he rolled over, and the can fell off the bunk, his finger could be amputated in a split second. Each of us sat there silently thinking about what that would look like, when compassion must've moved DeJesus to action. He offered to trade bunks with him. Lazada slept on the lower bunk where he could rest the large can on the floor until morning, and then right before wakeup call, the two of them switched bunks. When Sergeant Riley made his grand entrance, Lazada was settled back in the upper bunk on his back with the five-gallon can of oil resting on his stomach. "Been like that all night have you?" Riley scoffed.

"Yes Sergeant," Lazada answered. And that was pretty much the end of it except two of us worked for about ten minutes with a screwdriver to enlarge the hole and carefully free the lacerated finger.

CHAPTER NINE

"Manooch! What in the hell is that in your mouth? Are you chewing gum, boy?" Our platoon had just returned from the early morning run up Tank Hill and Sergeant Riley had us spacing ourselves at the proper intervals in order to come to attention.

"Yes Sergeant, that's what it is all right," I answered as the chuckles began to ripple through the platoon. I knew I was in for it. Chewing gum was taboo at all times except when off duty. And to make matters worse, my answer was a bit too sarcastic to go unnoticed.

"I'll tell you what, Manooch comma—Charles comma—the third. I want you to take that gum out of your stupid mouth, wet it down real good, and then stick it behind your right ear. You okay with that?"

"Yes Sergeant," I said as I spit the gum into my hand.

"Then do it!"

I did as ordered, but had to use considerable pressure and try several times to get the gum to stick. But it did, and there I stood with a gob of gum stuck to my head. *I remember thinking at the time, If this don't beat all!*

"Mahoney come up here. Each morning for the rest of this week I want you to come forward when ordered, remove the gum from behind Manooch's ear, put it in your mouth, chew it, and then tell me whether it's *old* gum or *new.* For Manooch's sake it had better be *old* gum. If we've got all that, then let's get down to Army training business."

That morning at chow I went over to the table where Mahoney was sitting to discuss some very serious business. "Look Mahoney, we can help each other out here. You don't want to put old gum in your mouth each morning do you? And, I don't want to be recycled back a week and have to resume training with a new Company. So why don't I place fresh gum behind my ear every morning, and you can chew it and tell Sergeant Riley that it's getting older and older." I looked over at him across the table and eagerly waited his response.

"Okay. For you I'll do it." He got up and walked away to dump his tray.

For the next several days Sergeant Riley, Mahoney, and I entertained our platoon each morning by reporting on the freshness of my damn chewing gum. I feel certain that the

Sergeant knew that I was using a fresh stick of gum every day, but his point was being made to his troops, and no one ever again chewed gum while on duty in Sergeant Riley's platoon.

It was about this same time in the cycle that we started our bayonet training. This was very intensive, and began with oral and film presentations made to us in a large auditorium. The introduction was intended to shock trainees—and it did. Big time.

"I want you men (that was the first time we'd been referred to as *men*) to pay very close attention to the screen." An old, grainy black and white film flickered a few seconds, and was then projected in front of us. "See those trenches there along the rise of that hill?" The Sergeant used a long pointer to direct attention to a spot on the screen.

We had to get our eyes accustomed to the dim light, but sure enough, we could see marks on the landscape that looked like scars.

"Well, there are U.S. Army soldiers in them and every single one of them is just like you. They *could* be you next time. Now, look down the hill. See all that blurred movement that goes clearly across the screen from one side to the other? That's bad shit there, and is actually a *horde* of Chinese infantry—several thousand of them. Some have guns, and others just have sticks as weapons. However, what every one of them has is a damn *bayonet* 'jerry rigged' out of something."

We sat there in utter silence as the Chinese advanced—ran—towards our soldiers who had begun to fire into them. It was mayhem, as the Chinese swarmed over and then literally hacked and stabbed their way *through* the defenders. This was no make believe play that we were seeing. It was obviously very real and left us feeling stunned, and at the same time very angry.

"You know how those two hundred poor bastards died? Over ninety percent of them were stabbed numerous times with bayonets by those slant-eyed, garlic-smelling bastards. Do you want to die like that?" the officer asked as he stared at the audience.

"No sir!" some of the fellows leaped to their feet and yelled, and most of us murmured in agreement.

"Then get the hell up off your butts and report for bayonet training. We're going to teach each one of you how to protect yourself, and then *you'll* be the ones doing the stabbing!" Thinking back to that day, he reminded me of General George Patton standing at parade rest in front of that huge American flag at the start of the movie *Patton*. He was addressing his troops

before a big battle, and told them: "You're going through the Germans like crap through a goose." He was expressing total distain for the enemy.

Like everything else we'd been through up to this point, bayonet training was extremely well organized. Instruction experience had been obtained by cycling thousands of trainees through the process. Then, there was that age-old command we had all heard since we were kids—"Fix Bayonets," which

allowed us to use that little funny looking thing that's under the end of the barrel of an M-1 rifle. The bayonet is slid into place over this implement and is then locked on.

On the offensive, the trainee basically springs forward from a modified crouch, and thrusts the blade into the enemy soldier's chest or mid section. He then draws the rifle back—extracting the blade, and is prepared to engage his next opponent. Recruits are also taught to slash an enemy diagonally from shoulder to hip, and how to use the bayonet to push aside an enemy's weapon. The commands "On Guard," "Thrust," "Slash," and "Parry" are used to direct these actions. The soldier can slash diagonally downward and then recoil in the opposite direction with a rifle "Butt Stroke." Slash down and butt stroke with the stock as the rifle is brought up. We went through these maneuvers hour after hour. The defensive, and then offensive, actions made me feel unbeatable in the "on guard" position while holding a rifle or strong stick or staff. If you've ever been through this training you know what I'm talking about.

The final component of our bayonet training was to complete the bayonet course. This is an exhaustive exercise where the trainee has to run, in full combat gear, through an obstacle course with his rifle and bayonet at the ready—to stab fabric-filled targets that pop up without warning along the course. We ran our course during a driving rain and had to run up and down the slippery trail—observed at each target by an instructor who graded our tactics.

Toby Lee and I started the course at about the same time.

"Manooch, are you as soaked as I am?" Toby said as he peered out like some pitiful woods critter from under his poncho

hood. Streams of water were running down his rain gear, and the only parts of him that were visible were his face with purple lips and the cold white hands that clutched his rifle. He looked as bad as he did when he got his hair cut at the Reception Station.

"I can only guess that I am. I'm damn wet that's for certain—these ponchos aren't worth a shit. I'm wetter on the inside that I am on the out." I pulled the poncho off, balled it up, and threw it next to a tree just as we started to run the course in the woods.

In addition to our rifles with fixed bayonets, each one of us was wearing a steel pot and had on a backpack with an entrenching tool attached. An entrenching tool is a small metal shovel, with a wood handle that folds up. During Basic we were constantly ordered out of ranks to form a hastily-established perimeter where we had to take our entrenching tools and dig furiously in the dirt to provide some protection from imagined incoming fire. The "dirt" was actually sand, and the top couple of inches of that was usually frozen—particularly early in the morning or in late afternoon. Entrenching tools could be useful in these instances, but when you were running they were always banging up against your back.

Our platoon in particular had extensive experience using entrenching tools—stemming primarily from an episode a couple of weeks before the bayonet training. It must have been around 2300 hours when we were rudely jarred from sleep by the lights being turned on, and shrill blasts from a whistle.

"Get the hell up right now. Put on your boots and fatigues and fall out in the street. Oh, and don't forget to bring your entrenching tools with you." Sergeant Riley was holding his damn whistle and shouting from the top of the stairs. We could hear our buddies down below scrambling out of their bunks at the same time. Less than five minutes later the entire platoon was standing in ranks in the cold night air in front of the barracks. We were still trying to wake up and wondering what in the world was happening.

Mahoney stood towering over me, but at the same time was trying to hunch down and get some protection from the wind. "What in the name of God are we doing out here at eleven o'clock at night? Don't the sons of bitches get enough out of us during our seventeen-hour work day?" he said with disgust.

"I don't know, but you can bet it isn't something good, "I said as I squinted from the wind with cold-induced tears running down my cheeks.

"Okay, *young ladies*!" Sergeant Riley shouted over the wind, which had picked up and was blowing from the North. "You're going on a little night detail. The place is close by, but you can bet we'll be there a while." Single file we trudged off between the barracks of the first and second platoons, went about fifty yards and halted in a vacant field. The ground was sandy and the soil so poor weeds weren't even growing in it. Not another soul could be seen outside at this hour.

"Circle up! Circle up! Form a big circle and space yourselves so your stretched out hands touch someone on either side. That's it. Move quickly. It's cold out here, and we don't want to be outside all night!"

"He thinks *he's* cold. He's got on his field jacket and cap and we're out here in our fatigues," I murmured more to myself than to anyone else. "I don't know what we're doing, but it ain't good," I said to Mahoney who was on my right side, and to Mills on my left.

"Ain't *good.* Hell. Dis is going to be a dilly. You wait and see," Mills said through his clenched, chattering teeth.

"Okay, start digging right in front of you and move toward the middle of the circle," Sergeant Riley ordered. It had become obvious now that we were digging a very large hole. The sound of forty-some rasping entrenching tools could be heard digging in the sand over our grunts, groans, coughs, and wheezes. We worked for more than an hour. First on top, and then gradually getting deeper into the hole, which had enlarged to about thirty feet in diameter and six feet deep.

"You can stop digging now. That's about as big as it needs to be." The Sergeant helped a couple of guys out, and it wasn't long before we were all out of the hole and staring down in it. We brushed the sand off our fatigues and wearily stumbled back to the barracks.

Around two o'clock in the morning I crawled back into my bunk and was slowly drifting off to sleep. Sand was still in my eyes.

"You know what bothers me?" a voice whispered in the darkness. It was Johnson and his bunk was two down from mine. "What if somebody comes casually walking along out there in the dark and falls in that hole? They'd break their damn neck that's what. It can't be left open out there. *Oh shee-itt* !"

"God damn it Johnson. You know good and damn well what the fuck is going to happen to that hole. We're going to go back out there and fill the damn thing in, that's what." McCoy had spoken the truth and we all knew it. Nobody went to sleep after that. We just lay there waiting for what we knew was coming.

At precisely 0300 we heard the sounds of someone stomping up the stairs—and we all knew who that "someone" was—Sergeant Riley. But it wasn't. When the lights were turned on there stood PFC Lee with a big smile on his face. PFC Lee was "permanent party," which meant he was now *duty station-assigned* to Fort Jackson, and could wear the infantry blue—designating his status as elevated above trainee. In essence, he could tell us what to do, and he was now happily ordering us to get up and report outside dressed as before. He turned around and almost gleefully skipped down the stairs.

"That little oriental fucker," Lazada said as he climbed out of his bunk and adjusted his swing easies. "Hell, he was willing to get up in the middle of the night just so he could order us back out to that damn hole." Everybody on the second floor of our barracks knew that PFC Lee was going to get his. It was only a matter of time.

"You probably wonder why I called all of you out here in the middle of the night like this," Mahoney said as we stared down into the gaping hole. There were a few chuckles, which momentarily replaced the moans and groans. It was cold as hell, and a light freezing rain was beginning to fall.

"If you want to get this over as soon as possible, then I suggest you start shoveling sand into that pit as quickly as you can." Sergeant Riley stood ramrod straight wearing his bloused fatigues, helmet liner, and poncho. Even under these circumstances he had that perfect military bearing about him. None of us were even wearing a "cover" (military slang for any type of issued headgear).

It was 0400 hundred when we finally crawled back into our bunks, just thirty minutes before we had to be in formation for the start of another day. There was no way we could fall to sleep in the brief time allocated.

CHAPTER TEN

The word around Company C was that we had several obstacles to deal with before we graduated from Basic Training. These included live hand grenade training, escape and evasion, battlefield gas training, nighttime river crossing, compass training, and the often-discussed infiltration course. It was not long before we started these more advanced areas of training in earnest.

First on the list just happened to be the hand grenade course. This exercise was held out in the open (an *inside* grenade course would certainly be disastrous) where we spent hours learning the anatomy of the grenade, how it is used, and for what purpose. We knew the latter had to be to blow hell out of your enemy—it's just a matter of how to do it safely for yourself and your buddies who happen to be close by.

The cadre also presented us with a brief history of the hand grenade. We learned that the word comes from the French *grenadoe*—probably derived from the shape of pomegranate fruit. Hand grenades were first used during the 700s by Byzantine soldiers. "Greek fire," which consisted of flammable materials, was placed into stone or ceramic jars that were hurled at the enemy. During the War Between the States both sides used grenades that had fuses. The older models of grenades in more modern times were often referred to as "pineapples," probably because they didn't have a smooth surface. They were dimpled, identifying the pieces of metal that would become shards when the grenade exploded. Each grenade was heavy, had a pin with a ring on it, and a metal lever. If the pin was pulled out by the ring—and the lever released—the grenade would explode within a few seconds. As long as the lever was depressed the pin could be replaced. Accomplishing that, my friend, would take a very steady hand.

The newer model, the M-61 grenade, is operated in the same fashion as the older ones, but has a smooth olive drab surface and the entire metal coating explodes into thousands of very small metal shards. The M-61 is about two and a half inches in diameter, contains five and a half ounces of Composition B high explosive, and weighs about one pound. We were told that an average soldier can throw one about one hundred and thirty

feet (I'd believe that when I saw it), producing a casualty radius of about forty-nine feet. However, the actual killing radius is much smaller.

Each trainee received individual instruction on how to hold the grenade, how to arm it, and how to throw it. The soldier had to stand perpendicular to the target, legs straight, feet apart, and with his arms bent—which allowed each hand, or fist, to be brought together and touch on the trainee's chest. Left-handed soldiers like me held the grenade in the left hand and looked over the right shoulder at the target. The pin was pulled by the right hand. Right-handed trainees did the opposite. This stage of the training involved dummy grenades, and was conducted on level ground. Several guys at a time were instructed individually while the rest of us watched and waited our turns. A couple of hours later, we were marched over to the live fire part of the grenade range. Loud "BOOMS" could be heard in the distance, and resulted in elevated interest on our part.

"I'm scared that I'll drop the damn thing," McCoy whispered over his shoulder as we marched along. "I've heard that some poor bastard is killed here each year trying to learn how to throw these things."

"McCoy, you're full of shit. This is *awesome.* They're going to show us how to kill more than one son of a bitch at a time—and blow hell out of 'em." You could see that Plank was getting into this and couldn't wait for his turn to do the real thing. I honest to God believe that his only regret was that the targets were not humans.

The live fire part of the training was conducted in an area protected by bunkers. One group had just completed their live fire component, and we filed into the bleachers as the others left. An NCO gave us a lecture on safety, and the consequences to everyone if something was to go wrong. The target looked to be a wooden pole driven into the ground and ringed with car tires. This blown-to-hell part of the range was about thirty yards or so in front of the throwing area. The trainee and instructor both stood in the earthen and reinforced bunker, with just enough headroom to see the target. A deep hole about ten inches in diameter had been dug into one side in the bottom of the bunker.

"That little hole there is called a 'grenade sump,'" the instructor explained as we all stood up to peer down at it. "If one of you dumb asses happens to drop your grenade then it's up to me to push the damn thing into that hole and get you and me out of the damn way before it explodes. If that happens, and we both get out of there alive, I pity the damn poor ass that dropped the grenade."

We sat there watching as each of us took his turn in the bunker and tried to throw the grenade as close to the target as possible. Watching that demonstration, it was obvious to us that members of our platoon would throw grenades in combat only as a last resort. I'd shoot every damn bullet I had before I'd pick up a grenade.

"You know what? Dem grenades could come in handy back home. Dey's a bar in Monks Corner where folks act real ugly to boys like me. Call me bad names and treats me like shit. I'd like to walk in there one Friday night when deys a bunch of dem in there all fat and juiced up. Roll one of dem grenades inside and then run out and slam de dow."

"Mills, sometimes you act as damn crazy as Plank does. At least you're *some* white. You don't know what it's like to be put down *all* the time," Ferguson Moody said as we headed back to the barracks.

"Isn't it better for you here in the Army where we're *all* in the same boat?" I said as we marched along.

"I guess so, Manooch. Let's forget about all that. Besides, I'm hungry enough to eat a horse."

The training schedule was really being squeezed now—with a lot more exercises to go, and only three-and-a-half weeks of Basic Training remaining. That night in the mess hall we tried to guess what the next exercise would be. "I think it'll be the river crossing," McCoy said.

"No, it is not," Mahoney casually remarked as he stabbed a piece of pork chop and popped it into his mouth. "The next training that we do will be learning to use the compass, and involves a nighttime exercise with aggressors and blank ammunition."

"Mahoney, how did you get to be so damn smart?"

McCoy replied as he stared at him.

"It has nothing to do with *intelligence*. All you have to do is wander by the bulletin board outside the Orderly Room, and read what's written there under 'Training Schedule'—it's all listed right there for anyone to see," Mahoney said as he got up from the table and headed for the door. McCoy and I got up, dumped our trays, and quickly caught up with him outside. We walked over to the Orderly Room. Without even thinking, I bent over and picked up a cigarette butt that was on the path and put it in my pocket. One less thing to "police" later.

The bulletin board was right there in plain view, and was sheltered by a small, shingled roof that sloped down on either side. The wood was painted white, and messages and other items of interest were pinned on the bulletin board. Sure enough, there was the typed "Training Schedule." "Well I will be damn," said McCoy. " I never paid any attention to the fucker."

The following morning the four platoons of Company C were marched out to the compass training area. As usual, we sat there perched on cold bleachers as our instructor explained in way too much detail how to use the M-1950 Lensatic Compass. Unfortunately, the man loved to hear the sound of his own voice, and thought that everything he said was very informative or was funny. This got old after a while, and even Mills and Lee had figured out that what should have taken thirty minutes to explain had taken most of an hour.

"Good God, I wish that son of a bitch would shut the hell up, and let us walk around awhile and warm up some."

"Moody, you're a pussy," Plank said. "This stuff is going to get good tonight when we can run around in the dark, shooting at everything and taking prisoners."

"I ain't even going to answer that," Ferguson Moody said as he shuddered and stuck both his hands under his armpits to get them warm.

The bleachers were emptied, and we were so stiff from sitting there for so long that we hobbled away like old men. The Company was divided into groups of about six trainees, and large topographical maps were spread out on wooden tables. Each group huddled around a table to receive instruction on compass use. Sergeants in charge pointed to locations on the maps. Soldiers had to locate them with the instruments and plot courses to reach each objective. Each trainee was required to "shoot an azimuth"—the angular distance from a fixed reference direction to a designated position—until he got it right.

The grand finale of the exercise was a nighttime maneuver. It required aggressors equipped with compasses, rifles, and blanks to locate a simulated headquarters or signal station. These "friendly" outposts were guarded by other trainees who were also armed. The aggressor forces were the ones actually being graded, because they had to rely on their newly acquired skills with compasses to locate positions.

You would have thought that Pvt. Plank had died and gone to heaven. That evening, before the exercise was to begin, he was running around getting pumped up—wearing a blue bandana around his head and with his face smeared by streaks of charcoal. All military decorum seemed have been temporarily dropped, and how Plank got all this stuff we never figured out.

What happened that night is as clear to me today as it was back in December of 1962. Four of us were assigned to one of the "friendly" stations—and one that could be too easily attacked by an "aggressor" squad. Our position was a wall tent set up on top of a hill surrounded by small turkey oaks which were about five or six feet tall. In other words, ours was a very typical spot on the Fort Jackson Military Reservation. The wind was blowing at a pretty fair clip, and the tent's canvas seemed to be breathing as it billowed out and then relaxed with each passing gust. Three of the fellows were inside the tent trying to keep warm, and I was taking my turn manning the M-60 machine gun outside.

You could hear Plank and his bunch of crazies stumbling around in the woods ten minutes before they got to us. I couldn't believe our luck. One of the aggressor teams was coming right for us, and that team was led by Pvt. "Gung Ho" Plank!

"You guys get the hell out here quick and be as quiet as you can," I whispered giving the only order I ever gave during Basic. By now the aggressors believed they had us pinned in so they started firing. The rifle fire was coming from one direction—easily located by the sound and by the muzzle flashes. Two of our guys arranged themselves in a defensive position in front of the tent as decoys and started firing blanks while Mahoney and I went to our right at the edge of the woods, where we crouched down to hide in the dark. Seconds later the aggressors charged out of the woods. The adrenaline rush, which had greatly enhanced their charge up the wooded hill, had since left them and they now stood panting for breath at the edge of the clearing.

"Come out here you fuckers, drop your weapons, and put your hands up!" Pvt. Plank yelled between gasps for air. "You are all now my prisoners." He was so exhausted that he could barely stand.

"Think again, Plank," I said as we stepped from the

darkness with our weapons pointed at them. Plank made a mad, unbalanced lunge at me, and I caught him off guard and pushed him flat on the ground. I quickly straddled his chest and thrust the M-60 across his throat. He couldn't stand that for long and soon surrendered. His buddies followed suit. When the referees came to our location for a battlefield evaluation they found that the "friendlies" had captured the aggressors, which was not the way it was supposed to happen. Pvt. Plank had at long last been in battle—but embarrassingly for him had lost his first skirmish. He hated me for that as long as I knew him.

CHAPTER ELEVEN

It was a happy day indeed when we received word that we would be given an extended pass over the Christmas holidays. Usually, graduates of Basic Training are given fourteen-day passes before being assigned to their next training duty station. Since it was so close to Christmas, we were given our fourteen days early, and I couldn't wait to call Genevieve and Mother and Dad. However, I was not the only one with the idea of calling, and lines had formed at the few pay telephone booths located close by our barracks. We had to use those phones, or none at all. I had to wait my turn to make one call, then go back to the end of the line and work my way back up to place the second call. By the time I got back to the barracks I was pumped up and ready to go on extended leave, and I felt as happy as I had for a long, long time.

Johnson was the only guy in our company I knew who lived near Raleigh, and he had a car parked off base. Two of his high school buddies had delivered it to a parking area just outside the main gate when he first reported to duty. The lot was strategically located to provide a means of personal transportation for those unable to keep their cars on Fort Jackson. Vehicles parked outside the gates didn't need to display the decals that were required for privately owned vehicles on Base. Blue decals designated officers; red enlisted personnel; and green or black for civilians that worked at Fort Jackson. Trainees in Basic were not entitled to a car decal.

"Johnson, do you mind if I bum a ride with you to Raleigh?" I asked that night in the latrine.

"Sure, I'd be glad to take you with me, but you don't mind helping with the gas do you?" he said as he sat on a closed toilet seat polishing his brass. (An E-1's brass is fairly simple—belt buckles and two circular infantry pins—one with crossed rifles, and the other with the initials "U.S." They are worn only on Class A uniforms—not fatigues). He had achieved a level of excellence and the small pile of brass literally glowed in the light as we spoke. Some guys polished their brass and then added a coating of varnish, which retained the desired luster. They did this at their own peril, because it was not allowed.

"Hell no. I'd be glad to pay you whatever it takes to get

away from here for a while."

Johnson explained that he'd carry me as fair as Clayton, and I could call and get a ride to Raleigh from there. Clayton's where he'd turn off to go to his home in Archer Lodge in Johnston County. He was like most country boys from North Carolina back then who loved souped up, fast automobiles. His was a beautiful black 1956 Chevy, and he proudly pointed it out to me the next morning as we walked through the parking lot. We had picked up our passes at the Orderly Room and were now as free as civilians for a couple of weeks. Most passes have a limit on the distance from Base that a soldier is allowed to travel, but that regulation was not in effect for extensive leaves given after Basic Training (or like ours, during Christmas).

Johnson and I were soon zipping through the countryside in the Chevy, and both of us were exhilaratingly soaking up our new freedom. We went through Columbia and were glad to see it disappear in the rear view mirror. Soon the traffic slowed to a trickle, and as we continued on we encountered a car only once in a while on that rural highway. The air felt so different and fresh that we would have rolled down the windows if it had not been so darn cold.

The South Carolina Highway Patrol had a reputation for being very aggressive with speeders—particularly with soldiers stationed at Fort Jackson. We had often talked about it, and had been cautioned by our cadre to be very careful when traveling off Base. Unfortunately for Johnson this reputation was played out when a tan and black cruiser emerged from between two tobacco barns and leaped onto the highway in a cloud of dust. We'd only been on the road for about forty-five minutes.

"Holy shit!" Johnson exclaimed as he immediately

started to slow and then drove off the blacktop to the right. We came to a stop on the shoulder of the road, and the trooper pulled in behind us with his lights still flashing. "Can you believe this? We've been through shit for weeks and weeks, and now when we just want to get home the damn police have to stop us and ruin the whole damn thing." Johnson was no "happy camper." I felt sorry for him but said nothing as I looked down at my hands folded in my lap. There was nothing I could do to help the situation.

"Can I see your driver's license, registration, and military pass?" the officer said as he bent over and placed his hands on the windowsill. You could tell that he wanted to look inside the car. "You fellows in Basic at Fort Jackson?"

"Yes sir," we both replied at the same time. It's funny how things work sometimes when you're under stress. Here both of us were, had nothing to hide, and yet were as nervous as cats. The officer had to feel it.

"Going back up to North Carolina for a few days? Says here about two weeks. That right?"

"Yes sir," Johnson said as he put his records back into his wallet. "He's going to Raleigh, and I'm going to a little place called Archer Lodge."

"*Archer Lodge*? I will be damn. I have a cousin on my mother's side that lives there. Her name is Wilson. Know any Wilsons there?"

"Yes sir, I sure do and *they're all real good folks*." Johnson was beginning to relax a little. It was contagious and I even managed a slight smile as I glanced over at the officer.

"It sure is a small world sometimes, Johnson. You-all be careful now. Slow it down a bit. And have a nice visit with your folks. You probably deserve it." With that he turned around and went back and got into his cruiser.

"Can you believe that? This has to be a very special day!" I said to Johnson as he started the engine.

"It's special all right. We're going home."

The rest of the trip was uneventful, and we arrived in Clayton in mid-afternoon. "Think you can get a ride home from here?" Johnson said as he slowed in front of a used car lot (used cars were not called "previously owned" back then—they were definitely *used,* and most sure looked it). "No trouble at all. There's a pay phone booth over there and I can use it to call my parents. Give me your phone number and I'll call you to get a ride back to Jackson—if that's okay with you." I really wanted to see Genevieve first, but she didn't have a car and I didn't want to bother Mr. and Mrs. Jeffreys with trying to arrange a ride for me.

Our telephone number in Raleigh was Temple 39545. There was no area code back then; you had to go through an operator to place long distance calls. And we used "TE" instead of "83" when identifying our phone numbers. Mother answered the phone when I called, and she seemed thrilled to hear my voice. "I can't wait to see you, and Chris has asked about you almost every day since you left." Chris was my little brother. He was eleven years younger than me, which made him eight at that time.

"I hope you're fixing something good for supper. It'll sure be nice to sit at the table taking all the time we want to." I was really going to enjoy that, as well as sleeping in my own bed with nobody getting me up at four-thirty in the morning.

"There's a ham baking in the oven, and I've made potato

salad and fixed deviled eggs to go with it. We're all looking forward to having you back home for a while. Her voice had that "how you must have changed" ring to it, and I knew that both of my parents were eagerly waiting to find out if I had settled down and was behaving more responsibly. "Tell me where you are and Dad will pick you up."

After relaying that information, I placed the phone on the receiver and stepped back outside where my duffle bag was sitting by the highway. The traffic was fairly heavy and I just stood there looking at all the people, and the variety of cars and trucks that passed by. Everything was so colorful. There was not one thing painted olive drab, and I don't believe that I have ever seen so many women and girls in the same place at the same time.

Thirty minutes later Dad drove up in his fawn brown 1961 Chevy. It was the same car that I had borrowed one night about a year earlier to drive a young lady back to Louisburg College. That was the fall semester of 1961—just one semester before my last at the college. The car stopped beside me and Dad jumped out and gave me a big hug. "You look good son—like you've lost some weight."

"Yes sir. They keep us on the move all the time, and there's certainly no snacking or eating between meals," I answered as I put my duffle bag on the back seat.

"Well, whatever they're doing it's working. And, I like this 'yes sir' business. The whole family is looking forward to seeing you, and we are eating at Mother Esta's on Sunday afternoon. You will be able to see everybody then."

The car pulled out onto Highway 70 and we headed towards Garner and then Raleigh. We passed right by the place

Harold Landis and I wrestled the gorilla, or whatever the damn thing was, when I was a senior at Broughton and Harold was a junior. It beat our butts that night, and it has to be a miracle that we escaped with our lives. If it hadn't been muzzled, the monkey would've ripped our throats out.

"Is it okay to invite Genevieve to eat with us? And too, I was wondering if you'd let me take your car tonight to go see her?"

"No problem with either. Your mother and I will be glad to see her again."

We drove right through Raleigh, taking the same highway down Wilmington Street, Hillsboro Street, and Glenwood Avenue until we reached Five Points. A few minutes later we drove up the hill on Fairview Road and I could see our house just before we turned onto Canterbury. The car had barely come to a stop before Mother and Chris came running out the front door towards us. I was smothered in hugs and kisses, which left no doubt that I had been missed and was loved. I was overcome with emotion, and had to turn my face so they couldn't see the tears on my cheeks.

"Chuck, you look fit and healthy," Mother said, with her arm around my back. "Come on in and put your things in your room. I know you want to change before supper." Chris had a big grin as he followed me down the hall.

I went into my old bedroom, which had been mine since I was three. It was obviously now a guest room. The wallpaper had been changed, and the toys of a young boy had been replaced with more grownup things. As I stood there I realized the room could never belong to me again.

CHAPTER TWELVE

I had barely checked out my old bedroom when some of the neighborhood boys I grew up with began to show up. Tommy Green lived across the street and John Ogburn lived next door, so they were the first ones to arrive. The three of us were back in the den trying to catch up on all that had happened during the past six weeks when Emerson Atkinson and Butch Royster joined us. All of them were in college. The schools were closed for the Christmas holidays, so we would obviously be seeing each other regularly over the next couple of weeks. I sat silently envying each one of them. How I would love to be home and in college like they were.

Before long the inevitable happened—our conversation got around to discussing sex. Of course Tommy was the one to get it started. "Chuck, I bet you haven't even *held, hugged, or kissed* a girl in six weeks, have you? I'd love to be a fly on the wall tonight when you and Genevieve get together," he said with a chuckle. The other guys started snickering, and I thought *what a bunch of dumb-asses, and it's none of their business anyway.* But I knew that it was. We always shared things together without getting too personal, if you know what I mean.

The truth is that the fellows in my barracks at Jackson talked frequently—and in great detail—about sex and what they were going to do when they were on leave. The description "horny as a double-dick billy goat" was often used to identify a person's level of sexual frustration. However, I don't want to get disrespectful here so that's where I'm going to leave it.

Soon after my friends had departed I called Genevieve, using the same number I had dialed for years—TE-40019. She answered on the first ring, and I was ashamed to realize that she had been waiting to hear from me all afternoon. We agreed that I would come to her house at seven o'clock. She also informed me that she had to baby sit for a couple that was renting one of her uncle's duplexes. Her aunt and uncle, Edley and May Wilder, lived in a two-story wooden house located on the left side of Poole Road—if you're going from Raleigh—and at the corner of Sunnybrook Road. Aunt May was Mrs. Jeffreys' sister. Uncle Edley was retired from the railroad, and he owned and maintained several rental duplexes in the immediate neighborhood.

"Why in the world would you agree to baby sit *tonight* of all nights?" I said with a groan. "We need to spend some time by ourselves, don't you think?" I said as I let out something like a disgruntled sigh, and for darn good reason.

"I sure do, but you can see me every day for two weeks, and I imagine that we can find some private time together. I owe Uncle Edley a favor, and he gave the couple my name as a reliable sitter. Uncle Edley's been awfully good about visiting with Papa, and even takes him out to just drive around and see things." Papa was her grandfather, who lived with the Jeffreys family until he died several years later.

That night I drove into the Jeffreys' driveway at quarter 'til seven. Mr. and Mrs. Jeffreys and their youngest daughter Ann greeted me with big hugs. Ann was about fourteen, and would become a cheerleader at Enloe—Raleigh's newest high school. She was as pretty as she could be. She seemed to always have a smooth, even tan and her smile would knock you over. As we were leaving, Mr. Jeffreys delivered his traditional warning to us, "you-all be particular now." We both knew what he meant by "particular." After he'd said that, he'd conclude with: "Be sure to be home by eleven."

I still remember what Genevieve was wearing that night. She had on a Scottish looking outfit that included a blue and green plaid kilt fastened in front with a large safety pin that was both practical and decorative. She had on a white blouse and white knee-socks worn with loafers. We stayed at the couple's house looking after two small children until the parents came home in time for me to get Genevieve back home before eleven. We still saved enough time to pull off and park for a few minutes.

Two nights later I had an evening out with the guys. Charles Debrito, Bill Farrior and I went to a small tavern at Five Points, which was located about two doors up Glenwood Avenue from the old Colony Theater towards the Fairview Road intersection. Eighteen and nineteen-year-olds didn't have any trouble being served beer back then, so we ordered a pitcher of dark beer. Back in the late fifties and early sixties teenagers drank beer like they do today. They drove around in cars while drinking, and yet I can't recall a single incident when a teenager was charged with driving while "under the influence." In the tavern, music—mostly 50s-early 60s rock-and-roll—was coming from somewhere, and provided a pleasant background for drinking beer and catching up on all the news around town. Buddy and Michael Cayton came in after a while and plopped down at our table. They joined in our conversations that were getting louder and more nonsensical as the hours ticked by. Buddy had played football with me at Broughton. Actually, he played and I normally sat during the games. He's a year younger than I am. Michael is the same age as me, and was in my high school class of 1961. The five of us stayed at the tavern until the two a.m. closing time, and then worked our ways back to our parents' homes.

One of the things I often thought about and missed most while on active duty was fishing in Mr. Jeffreys' pond. I fished with him in the spring and summer, catching bluegills, shellcrackers, largemouth bass, and warmouth. The pond was small, about half an acre, and it didn't take long for me to learn the best spots to fish and what baits to use. Most times I fished from the bank with a cane pole or fly rod, but Mr. Jeffreys had a small aluminum pram that we used on occasions. He enjoyed operating the trolling motor and positioning the boat in just the right spots for me to fish—usually in shallow water next to the shore and adjacent to aquatic vegetation where the fish were bedding.

Unfortunately, December was not a good time to fish in a small private pond—in Wake County, anyway. However, Mr. Jeffreys' land, which extended well beyond the pond, provided excellent opportunities for hunting small game. The shotgun I hunted with back then was a good gun, but was not very practical for field and woods hunting because it was heavy and had long barrels. It was a 12-gauge double barrel *Fox* that I purchased from Hill's in Raleigh for forty-five dollars. I took it, some shells, and my hunting coat, which included a large pocket in the back for game, on a hunting trip one afternoon on the Jeffreys' farm. Camouflaged hunting apparel had not yet come into its own, so most hunting outfits were a cured tobacco-brown (you know the color) and were made of a canvas-like material.

I started the hunt by following the path that crossed the pond dam and then worked my way up a slight rise to an open field, which was about fifty by twenty yards in size. It was located generally where the I-440 Outer Loop is today—between Samaria Baptist Church and Dowling Road, and about three hundred yards southeast of Poole Road. The field had not been plowed or planted for several seasons, so it was overgrown with patches of blackberry, dog fennel, and ragweed—an excellent place to flush rabbits and quail. I poked around the blackberry thickets for a few minutes, startled a rabbit from its hiding place, and shot it before it ran too far. It was a good shot and I rolled him up dead as a door nail. Unlike some other upland small game species, rabbits tend to be more delicate, and only one or two pellets from a shotgun load usually will be enough to kill one.

Having no more luck in the field, I entered the woods at the far end, went a short distance and sat under a large red oak tree. It stood out because it was by far the largest, and also because most of the trees in the immediate vicinity were small pines and cedars. An old, rusty barbwire fence was barely nailed to the oak, and I sat leaning against the tree next to the wire. I remained absolutely still and after a while I could hear birds and squirrels rummaging on the ground through the fallen leaves. It was the time of day when animals make their last attempts to find something to eat before turning in for the night.

Squirrels usually mate in the spring and fall, and not in December. However, at least two of them were feeling frisky that evening, and they sounded like horses chasing each other as they ran through the woods. I remained motionless and seconds later the first squirrel ran over a little rise and the second was close behind. They were following the barbwire boundary. I quickly drew a bead, fired once, then a second time, killed both, and stuffed them in my game bag.

The sun was sinking low over the trees at my end of the field when I emerged from the woods. Birds crisscrossed in front of me in the dim light as they flew in and out of the old field, feeding on seeds and insects that would comprise their last meals of the day. I was standing there taking in this quiet setting, when I heard the whistling sounds of doves' wings overhead. The sounds are actually more "squeak-like" than whistling. I've been told that the sound comes from the wind as it passes over special feathers on the wings. The birds were returning to their roosts, which were located in cedar trees clumped close to where I had first entered the woods. Since it was not yet sundown, shooting was still legal, and would be for another twenty minutes or so. I positioned myself about ten yards in front of the cedars and waited. The birds flew in fast, and their aerobatics would have made fighter pilots proud. My shotgun had to be shouldered quickly, and swung from a variety of angles to increase the probability of delivering a good shot pattern. But I did it, then retrieved the dead doves and headed for the house up the hill.

"What did you shoot?" was the question asked when I arrived back in the Jeffreys' backyard. Genevieve and Alice Gayle's children, Andy and Gayle, all watched as I emptied the contents of my game bag—a rabbit, two squirrels, and six doves onto the dirt path. "Not bad for a two-hour hunt," I proudly said.

"What are you going to do with them?" Andy asked as he timidly poked at the rabbit with a twig.

"We're all going to *eat them*. That's what. When you shoot something you at least ought to eat it so it died for some reason."

"You might, but I'm not. So there should *plenty* for you," Genevieve replied as she turned her nose up at my furry and feathered trophies, and walked up the back porch steps.

"Girls! They are no fun," Andy said as he reexamined the rabbit's furry feet, and held up a dove by the tips of its wings like it was still flying.

"Well. Sometimes they are." I said as I picked up the game and went around to one side of the barn where a garden hose was attached. The cleaning responsibility was all mine.

I stood there by myself for a few minutes looking at the pond, field and woods as darkness approached. It was quiet and peaceful, which combined with the natural beauty, created a scene that has stayed with me. I wondered then how long all this would remain as it was that day, considering Raleigh's growth into surrounding rural areas. Sadly, not very long.

CHAPTER THIRTEEN

The afternoon of Christmas Eve arrived and my parents made arrangements for us to go over to Dick and Edna Hines Parsons for their annual neighborhood party. The Parsons lived on Runnymeade Road at the corner of Lake Boone Trail at the upstream end of Boone's Pond. Their two children, Richard (we called him "Skeeter") and Julie would be there as well. It was always an enjoyable time and I looked forward to it every Christmas. When I was younger I had to sneak back into the kitchen to help Dick make the eggnog—tasting it of course as we went along. *Somebody had to do it.* Understandably, this could take us a while. However, on this occasion I was allowed to join the adults and sip eggnog in the dining room—right out in the open—and was also permitted to select a mixed drink from the bar. It was pleasant being in the company of so many civilians for a change, discussing topics that that were varied, and that didn't pertain to the U.S. Army or Fort Jackson. Many of those at the party had children who were either in college or were planning to go there, and I wanted to learn more about it. Helen and J.M. Jenrette's two children, Helen and Buddy, were there. I was developing a very strong desire to join my young friends in their academic pursuits as soon as I returned home, and had earned the opportunity.

After spending a couple of hours at the party we said our goodbyes, and Genevieve, Chris, Mother, Dad, and I left to go to Annie Laurie's and Uncle Truman's to continue our traditional Christmas Eve celebration. The drive from the Parsons home to Harvey Street was beautiful. Homes with windows and doors elaborately decorated looked like something out of a magazine. My great-aunt and great-uncle lived at 910 Harvey Street in a stone French Norman house they built back in the late thirties. They always invited the family over for supper on December 24th. This was the first time all of my family had gotten together since I returned home from the military. It proved to be a joyful reunion with my grandparents Esta Callahan, Loulie and Big Daddy Manooch, and my other great aunt Maude Stewart and her husband Roger. Estie's husband J.W. Callahan (my granddaddy) had died just before I enlisted.

The setting on this evening was as I had remembered from Christmases past. The slate walkway was illuminated with candles that were within glass chimneys held aloft by metal poles. There was a large evergreen wreath with red bow on the front door. And a warm fire greeted us in the living room. A small artificial tree stood on a table off to one side of the fireplace. I often wondered why a house as beautiful as this one, with its large high ceiling foyer and spiral staircase, would be decorated with a puny artificial tree.

"How does it feel to be home, Chuck?" Uncle Stewart was standing by the fireplace and spoke to me between sips of warm apple cider, the aroma of which filled the room.

"Great. In fact I wish I didn't have to go back."

"Come on, Chuck. You know that's not true. The military is exactly what you need now until you decide what you want to do with your life." Dad had walked over to join the conversation.

"Yes sir, you're right," I responded to Dad's statement. "Fort Jackson will certainly help define my future." *I knew good and darn well what I was going to do when I got back home in May. I was going to straighten up, and go to college like the other guys I knew if it meant applying to a hundred schools. Dad would probably be disappointed if I didn't choose the military as a career, but he'd get past that if I did well in college. After all, he didn't go to work with the Seaboard Railroad as his father did.*

Annie Laurie had live-in domestic help as long as I could remember. She and Uncle Truman had a housekeeper and cook named Janie, who resided in an over-the-garage apartment from the time I was born until I returned from active duty. However, Janie quit and moved out when her grown children in Washington, DC told her it was demeaning to work in the house of white people. I often thought later, after the race mess had temporarily settled down, whether Janie had regretted her decision to leave. I know I missed her. She was as sweet and kind as she could be—and we all loved her.

Our Christmas Eve supper was always formal, at least to me, and the menu was fancier than I was used to. Henry, a Negro gentleman who had worked for Annie Laurie and her sisters for years doing odd jobs and yard work was now working in the kitchen. He came in to announce that dinner would be served shortly. Once we had taken our places at the table, we were served half of a baked acorn squash that contained a mixture of butter, brown sugar, and walnuts. The brown sugar and butter combined to form an amber-colored syrup, and the walnuts floated in it. Next we had a mixed green salad that contained a lot more than iceberg lettuce, and was served with a raspberry-vinegar dressing. Under the table, and in front of Annie Laurie's chair, was a small button sticking through the carpet that she could press with her foot to notify the servers in the kitchen that she was ready for the next course. Dishes were prepped in the butler's pantry, which was just off the dining room. We were offered a choice of entrees, including roasted Cornish hens basted with apricot jelly, and Virginia baked ham. Both were served with sliced crabapple, asparagus covered with a yellow cream sauce, small roasted potatoes, and snap beans with almonds mixed in. Henry cut the ham on the sideboard, and the slices were very thin and cut almost horizontally from one side of the ham.

Down at the far end of the table Uncle Truman was playing with his asparagus with a fork, and was trying to cut one in half. "This stuff is practically raw—it needs to cook some more."

"Mr. Williams," Annie Laurie replied. "That's how

you're supposed to cook some vegetables. They say it's better for you and vegetables are served that way by top-notch chefs in the fancier restaurants."

"Darn with what they say is 'good for you.' Can you imagine sitting down to a plate of collards that have been cooked for five minutes? I want my vegetables *cooked* like they're supposed to be—and not to *impress* somebody."

"I'm with Truman on that," Aunt Maude chimed in as she helped to pass the plate of rigid asparagus.

"Give the plate to Chuck. He'll eat two of anything," Mother chuckled. She was right.

Other than the debate about how to cook vegetables, the conversation had been rather subdued up to this point, as mouths were too full to utter a word. However, it picked up as we were served coffee and dessert.

"How does this food compare with what you're used to in the Army?" asked Annie Laurie from her end of the table.

"We never get anything to eat that's this fancy, and we can't just sit around and talk when we're through eating. Taking time to just sit here and leisurely talk has to rank right up there with the excellent food." *There were, however, at least three very favorable aspects of Army mess hall dining. First, the food was typically good. Army cooks had been to school and learned how to prepare food for the troops. The cooks took pride in what they served, and were graded periodically on sanitation, service, and quality. Second, there was plenty of it. Every soldier received ample servings. It would be difficult imaging anyone leaving a mess hall hungry. And last, the mess hall and kitchen were extremely clean. This could easily be accomplished since each company had approximately two hundred soldiers that could be ordered to KP duty.*

Completely satiated, we gathered back in the living room. While we visited there I had the chance to think back over previous Christmas Eves in this house. When I was young—four or five—I was given one present to open before Christmas day. It was always a toy that I wanted, and I had often gone with my aunt and uncle to help pick it out. One of the first things that I remember, even before the special presents, was a small metal stand that was intended to hold a tree or cake. We never used it for those purposes, but it was brought out for me to play with. It was green and white and could be wound up to play *Silent Night.* I unfortunately recall another time when I had excused myself from the table and had gone back into the living room where I found a bag of dried apricots, ate them all, and got very sick. Every time I see dried fruit today I have to fight the urge to gag.

On Christmas day the family traditional breakfast was at our house and our close relatives joined in the festivities. We ate, opened presents, and that was it. Another Christmas had come and gone. Soon afterwards, only a few days remained until I had to return to active duty. Fort Jackson seemed like a million miles away to me right then. However, as part of my Christmas present, Dad had planned a hunting trip to Lake Mattamuskeet. I knew nothing about it. He, Harry Moore, Owen Walker, J.M. Jenrette and Dick Parsons often hunted geese and ducks there, and as a child I carefully studied the photographs of the smiling faces of Dad and his friends with their shotguns and dead waterfowl. The dead birds were piled in front of the group of hunters so it was impossible to tell who had shot what. Tim Timberlake and

I hunted at Lake Mattamuskeet one time when I was in high school and he was home from a military school. That had been a very enjoyable trip and I often thought about going back. This seemed to be my chance.

Dad made arrangements for us to hunt on private property that was adjacent to the Mattamuskeet Refuge on the southeast side of the lake. The property belonged to a Mr. Snow who owned the Beaufort Equipment Company in Washington, NC. A friend of Dad's, Ed Nohldemeyer, agreed to be our guide. Ed was a pilot part time for the North Carolina National Guard and full time for the U.S. Fish and Wildlife Service, stationed in Washington. His work area with the Service included the large lakes located on the Albemarle-Pamlico Peninsula, and he spent many hours flying the area to assess waterfowl populations, and at the same time aid the law enforcement fellows capture illegal hunters.

Dad and I left Raleigh early one morning and made the two-hour drive to Washington where we picked Ed up. Another hour took us through the little towns of Pantego and Belhaven, and then to the lake, where we parked off the road at the entrance to Mr. Snow's property. The drive through this part of rural North Carolina is beautiful to those of us who like less populated areas, and the wild things that live there. Facing east at Mr. Snow's gate, the town of Englehard is about four miles away. Woods and wetlands are located on the left side of the highway, and large fields with old two-story farmhouses and their outbuildings are on the right. Geese fly off the Refuge to feed on grains they scavenge in the fields. Farmers are often subsidized by the government if they follow farming practices that are somewhat less profitable for them, but at the same time allow food to remain in their fields for the geese and ducks. On this trip in early 1963 there was seldom a time when flocks of geese could not be seen and heard in the sky. The over- wintering Canada goose population was near an all-time high during this season in Hyde County.

We unloaded our gear, which seemed excessive for just an overnight stay, and stored it in an old army three-quarter-ton truck. It had seen better days, and had paint peeling off. It was a good thing that we had a vehicle of that type because the highway was about a mile from the lake. We had to go through the woods following a trail that was nothing more than two muddy ruts. The mud was black, had a sulfur or rotten egg smell to it, and it obviously had not been disturbed for a long time. The woods were thick on both sides and so close to the trail that the limbs blocked out much of the light from overhead. We could hear the "*Ka-Ronk*" calls of the geese off somewhere in the distance but were unable to see them through the trees.

"It's not far now," Ed told us as we plowed on through the mud. The windshield wipers were going all out to allow at least a glimpse of the road, but they were dry rotted and did little more than smear the grime on the windshield. "Mr. Snow has a small cabin that he said we could use, and once we're settled in we can take a look around the property."

"It was nice of him to let us hunt here," I said as I tried to concentrate on something other than the pitching and rocking truck. I was close to being sick from all the motion. The hamburger and fries that I had for lunch in Belhaven were not helping the situation one bit, and in fact were trying to make an upward escape.

Just as I was about to lean over the side to get some relief, we drove into a clearing and the cabin came into view. It was small and looked to be a one-room deal with a porch on the front. There were two or three windows and a brick chimney, which gave the little shack a homey appearance. *Only a hunter could love this place I thought.* The porch wasn't screened, and I wondered what it would be like sitting there in the late spring or summer with no protection from the hordes of mosquitoes that were bound to emerge from all the standing water which made the few scattered elevated places look like islands.

Once we parked and were inside the cabin, Ed explained that we had to place black plastic over all the windows before it got dark. "There will be thousands of geese on the ground not four hundred yards from where we are now, and I don't want to spook them by showing too much light when it's dark. They're smart as heck and have sentries always on alert for trouble."

"You mean that geese are afraid of the light?" I asked Ed.

"No. Not in general. But they are wary of light at night coming from an unexpected place—like from this cabin in the middle of a swamp. We want them to stay close by overnight, so when they fly out to feed in the morning they'll come by low—just over the trees."

It was getting late when we went out to get "the lay of the land." We first inspected two ponds that were located near the cabin. The ponds were bordered by freshwater marsh grass and myrtle shrubs. A hunting blind constructed of wood was on each pond and faced several duck decoys that floated on the water's surface. "These ponds can be good for hunting ducks sometimes, either waiting in a blind or jump shooting them off the water," Ed told us. I didn't pay much attention to the ponds because I was there to hunt geese, not ducks. We surveyed the boundary of the property to get a "feel for it" and then returned to our cabin. It was getting dark when we got back, and our homestead looked awfully small surrounded by all those trees rising up into the darkness.

After we ate, and before we turned in for the night, Mr. Nohldemeyer gave us instructions for tomorrow's hunt. "We'll want to use 'high brass' shells—either No. 4s or No.2s. Magnum loads. The geese will probably be flying close to the outer killing range of a shotgun so you'll want to put a good number of shot in them. The down feathers on the breast are soft, but arranged to provide padding, thus protection. It's sometimes difficult for the shot to penetrate."

"Where's the best place to shoot a goose?" I asked as I took shells out of a box and placed them one at a time in the slots of my hunting vest.

"Anywhere really. But the best way to bring down a flying goose is to quickly move your gun with him. Start with the tail and move the barrel through the goose until you've crossed right in front of the beak. You won't have much time. Then fire. The closer he is to you, the less you have to lead him." Ed was already climbing into his sleeping bag. "Lights out."

Just like at Fort Jackson, I was awakened in the morning long before I wanted to be. I peered out through a crack in the shade. It was pitch black outside and I tried to make sense of where I was and what I was doing. Ed was the first one up.

" Didn't hear much honking at all last night. That's a good sign. When they're flying at night and vocal it means they are feeding and moving to and from the fields across the highway—usually on a full moon. If they've stayed put—then they'll be hungry and ready to move at daybreak." Mr. Nohldemeyer was still teaching his students.

"I think I'll stay here. You-all can go out at this God-forsaken hour and hunt all you want," Dad said with a sleepy smile.

He did get up though. We all dressed, and the three of us slowly felt our way in the darkness until we came to the boundary of Mr. Snow's property. It extended to the Refuge and was separated from the government land by a water-filled canal with tall pines and scrub myrtles on both sides. I didn't know it then but most of our shots that day would be directed just above those pines. We hunted all day and killed four geese. I remember shooting two, and I think now that they let me have the best shots.

It was a good time spent with my dad, and the memories of that day linger still. Particularly the feeling of the cold wind on my face, the gray overcast sky, and the prevailing smell of pines and myrtles. All the honking that was with us for those two days made an impression on me too. Days after I returned to my barracks I lay awake and from far off came the wonderful wild calls of flocks of Canada geese as they talked to each other and winged their way through my mind.

CHAPTER FOURTEEN

Saying goodbye to my family and friends was especially hard. However, the situation was made more bearable because I envisioned myself as a soldier going off to the wars to protect Raleigh and its citizens from invading armies. A "hero" in the eyes of those left crying and waving along the roadside. In truth, I would be almost as safe as those going back to school after Christmas break. And not many tears would be shed for me, certainly not along the highway. Genevieve was the notable exception. She and I were becoming more comfortable with each other. That may sound curt and unromantic, but from my perspective it implied a healthy relationship—one both of us had grown to cherish. *It might*, I thought as Johnson and I left Clayton, *eventually lead to something more permanent.*

I slept most of the way back to Fort Jackson, and woke up only when Johnson's car pulled into the parking lot outside the gate. We were both wearing our winter dress greens and I took a cloth and quickly buffed the toes of my shoes and the bill of my cap before stepping out of the Chevy.

"How do I look?" I asked as I walked around to the driver's side.

"Smart, Manooch. You look just fine. I know I look okay. Let's get back and check in before supper. I've missed that damn mess hall." I couldn't tell if he was kidding or not. We entered the Orderly Room to turn our passes in to the Sergeant on duty, who was sitting at a desk and gave the impression that he was intently reading through some papers. He looked up and grinned when we approached—like he'd been waiting for us in particular.

"You know, Manooch. I learned something about you over the holidays. I understand your daddy's a Major. That right?"

"Yes Sergeant. He is. He's been in the Army ever since the Second World War when he was a bomber pilot." *Damn I thought. Is Sergeant Riley trying to be nice*? He sure as heck wasn't.

"I hoped you'd gone AWOL Manooch. How'd that look? A Major's son running off and too *chicken* to report back to duty?" Riley smiled as he said this. At that moment I hated his guts.

"I'm sorry," I said as I returned his casual smirk, "but I'm still here, Sergeant, and ready to roll." Johnson and I left the Orderly Room and intentionally let the door close quietly behind us. Outside the smell of coal smoke hung heavy in the air and left no doubt that we were back at Fort Jackson. It cast an oppressive light gray blanket over the Company area.

We had about three weeks of Basic Training left in our cycle, and we had a hell of a lot of training to do yet before we graduated. Talk throughout the Company once again centered around the subject of what field exercise we'd be doing next.

"Well, look what the cat dragged in," McCoy said sarcastically as Pvt. Plank sauntered over to our little gathering and sat down. As usual, we were clustered beside Moody's bunk as we attempted to solve the truly important problems facing the world. We did this often.

"You guys plotting against the *white* folks?" Plank said as he came over and eased down on a footlocker.

"Naw. Dey's too many of 'em,'" Mills replied. "And, in case you hadn't noticed some of us *is* white, dick head."

"Some. Right. But you're sure as hell *not one of them* Mills, you kinky headed coon!" Plank's voice rose as he stood up and stared in Mill's face. The two had to be separated before we all got into trouble. If an NCO heard us we'd all be on work detail for a week. Things settled down as quickly as they'd flared up.

"Anyway, I just wanted to tell you smartasses that this week we're going to do the Infiltration Course. I pray like hell that one of you gets so *scared* that he stands up and gets stitched across the belly with .30 caliber bullets. Then we'll see who's tough," Plank said as he headed for the stairs.

"Damn. What a great guy. And just think. He's on *our* side," Mahoney said as he broke the seal on a new deck of cards. "Anybody for a game of poker?"

Sure enough, the infiltration course was next on the agenda, and the entire Company assembled for instructions before marching out to the training area. We had all heard what has to be one of the most often quoted stories in the U.S. Army. It was the tale about the soldier who was crawling through the course when he encountered a rattlesnake. He was so frightened that he jumped up and was shot. We had that episode dancing around in our minds when Captain Reynolds, the Company Commander, provided some insight on what we were in for.

"The infiltration course is as close as you'll get to combat without being there. You will crawl the one hundred yard course three times—one of them at night. Live machinegun fire will be directed a couple of feet over your head at all times. Therefore, it is imperative that you remain as flat on the ground as possible as you encounter barbwire, bunkers, and trenches. *Do not raise up*—even on your knees. You could be killed. Remember to turn over on your back when you come to the barbwire. There are two types of wire—strands that are easy to get under, and concertina, which is rolled, and held into the ground by large metal screw-looking anchors. Hold your rifle in the "at arms" position when you're on your back, and hold it so the tip of the barrel extends to the top of your head. Use the rifle to probe the wire, lift it and then guide it so you're able to wiggle under. Your experience will be complicated by simulated artillery and mortar explosions, which will greet you along the course. It's very important that you don't freeze. DO NOT get pinned down, or you're no use to

your fellow soldiers. Keep moving and continue to gain ground. Remember *stay down*—and good luck."

"Jesus save us," Mumford said as he spit out his wad of tobacco. "I was hoping I could get out of this one, but even the Company Firemen have to go through it."

"You'll be a better man for it," McCoy said with a wink.

As stated, the course covered a distance of one hundred yards—all of it wet sand when we traversed it—and was uphill at an incline of about ten degrees. It was raining pretty hard when we went through it the first time, and sand stubbornly clung to our uniforms.

I'll never forget the night exercise. During the day we were certainly aware of the machinegun fire by the *rat-a-tat* sounds of the guns, and by the occasional whine of the bullets as they zipped past. However, when it was dark the "tracers" that were spaced at about every fifth round, lit up the night as they sped just above us, and then ricocheted off rocks and trees down in the bottom several hundred yards in the distance. Illuminated like that, the bullets were seen as real and very close and personal to us. We had a tendency to get mesmerized and stare at the tracer rounds, but we struggled to keep moving like our Company Commander had told us. During the night exercise troops were also exposed to flares, which were shot high into the night sky and would slowly drift back down to the ground. We were instructed to remain perfectly still when a flare was overhead. *I've wondered since that in actual combat it might be best to dash for cover rather than freeze there out in the open with every thing as lit up as a car dealership. I guess what the cadre were trying to stress is that movement draws attention.*

I made my way up the course relatively easily, and was one of the first at the top as I slithered past the fixed machinegun positions. I stood up and looked back at the course with both a sense of pride and relief, and realized it did resemble a real battle situation. Mahoney was right behind me.

"Damn, when those bunker-things exploded that really put the 'giddy-up' in my behind." He was smiling, and like me was proud to have finished the course, and ahead of the other guys.

"If we got through that okay, I guess we can handle the rest," I replied as I stomped the wet sand off my boots.

Part of the "rest" was gas mask training and that included a gas contact chamber. The mask we used was a U.S. Army M-9/M-9A1 Protective Gas Mask brought into service around 1951. It had a 60mm drum located on the left side. Someone wearing one of these things looks like the feature monster in a horror movie. More precisely, with its triangle-shaped, clear plastic covered eyeholes, the wearer resembles a huge alien insect. The mask has four straps—two that go over the head, and two

on either side—and a knob-looking thing over the mouth. We received an hour's instruction on the types of gas used throughout history, the physical and physiological effects of each, and how to respond to an attack—including the proper way to put the masks on. When the alarm "Gas!" is given, the soldier whips the mask out of its case, places his two thumbs inside the mask to spread it open, puts it over his head, positions it in place, and properly secures the straps. The "All clear!" command results in the reverse of the sequence.

Once we had completed our dry runs it was time to be exposed to the real thing—disabling gas.

"We are going to do this at the squad level. When your group is called, your squad leader will take you into that little building over there." The NCO in charge directed our attention to a small cinder block hut that had a door but no windows.

"Shit. You mean we have to go in there and get gassed?" Lazada moaned as he adjusted his gas mask container. The actual masks and containers gave off a fresh rubbery smell and were new issue—something we seldom encountered in the Army. We were all glad that the masks were new. We sure didn't want to wear masks that had already been exposed to gas.

"I heard they use *real* poisonous gas in there too."

"Lazada. You know dey can't do dat! Hell, somebody could die in there." As usual, Mills was getting highly agitated, and was causing some of the fellows to get nervous. Our squad filed into the hut, and we were ordered to sit on the concrete floor around something that looked like a heating element with a pan of water on top.

"You guys pay very close attention. What we're getting ready to learn could save your life. When I yell 'Gas!' I'm going to drop this tablet in that pan and it will produce tear gas. You must hold your breath until you put your gas masks on. Make sure they're properly sealed—or they won't do you a damn bit of good." The Sergeant dropped the tablet into the water.

A thick cloud of the non-lethal gas began to rise from the pan—reducing our visibility to a couple of feet. I recall thinking that *we resemble a bunch of Indians in a teepee squatting around a pile of hot rocks with water poured on them.*

"Everyone up! Get out the door!" The Sergeant commanded.

We all stumbled outside, and were ordered to vacate the

area by running about fifty yards. *It was during this exercise that my respect for firemen with gas masks on and running up stairs took a quantum leap. It takes a real man to do that. You want to know what a hero looks like? Well look into the eyes of a fireman wearing a gas mask who's run up twenty flights of stairs to rescue a child from a burning building.*

"God almighty. I think I'm going to pass out if we don't stop," Johnson said as he gasped for air through his mask.

"Yep." Is all that I could manage as a response.

"Seems like I've been here before." Mahoney was trying to be funny. Occasionally he was, but he could pick the darndest times. We were once again inside the hut and were too exhausted and nervous to fully appreciate his wit at the moment. This next phase of the training would have us exposed to a deadly gas.

"Look, Mahoney. We know you're trying to give us a lift, but we could all die in here this time around. Come on." I kept taking my mask in and out of the container to examine it for flaws. "Hell. Chlorine gas is bad shit. You heard what they told us about it. Several whiffs—boom—you're dead."

As if on cue, the Sergeant came in and squatted down just outside our little circle. "He's right. This stuff can kill you. But it won't if you do exactly as you're told and like you've been trained today." He reached in a bag and took out a small white tablet.

"When dissolved, this tablet will release free chlorine gas, which will then rapidly diffuse throughout this room. This is very serious. You *must* hold your breath and then calmly put your gas mask in place and *seal* it. If you do this you will be okay. Let's do it."

"Gas!"

I got the slightest sensation of chlorine just before my mask had completely sealed. It reminded me of summers at Hayes Barton Swimming Pool.

CHAPTER FIFTEEN

"People see us everywhere…
They think you really care…
But myself I can't deceive…
I know it's only make be-lieve…
Dah-a- Dah- Dah- Dah."

The sultry, rich voice of Conway Twitty filled the barracks bay, and each of us in his own way thought about people and things very far removed from Fort Jackson, South Carolina. Evidently, someone had snuck a portable radio into the barracks. Until it was discovered it would provide evening interludes, which would at least *mentally* remove us from our situation—if for only a short period of time. If we were careful, we may not be found out for a few days, and could receive evening serenades by Twitty, Elvis, and other popular crooners.

The days were going by, and we had little more than a week to complete Basic. Several obstacles stood in our way before we graduated, and we wanted to buckle down and complete our cycle on schedule. The alternative was unthinkable.

Our next tactical exercise was conducted at night, and we were transported by trucks all the way to the other side of the Base to the training area. We disembarked and realized that we were standing near water. A cold wind was blowing off of it and stung our faces.

Someone noticed movement along the far shore, but it was too dark to make it out. Our "Nighttime River Crossing," as this little adventure was referred to, was being made at what I would call an *enclosed body of water* like a lake, not a river. But I could've been wrong—seeing it only at night. The inflatable boats that would ferry us across to the other side were made of black rubber, and a squad-size unit would cross over in each boat.

"I haven't told anybody this Manooch, but I can't swim." McCoy was obviously afraid, and for good reason, but he didn't reveal his discomfort to anyone else in the boat. And to his credit, he didn't hesitate to take his place on a seat just behind me on the right side. We were sitting with our life preservers on, close to the bow, when we shoved off.

The idea of this assault was to make as little noise as possible. The "enemy force" on the other side was instructed not to fire unless the order was given—and then only if we could be heard crossing the water. This was very subjective, but was intended to give us at least a chance to accomplish our mission. The NCOs had given us instructions on how to use the black short-handled paddles to "feather" the strokes and thereby reduce splashing. I couldn't believe how silently sixteen rubber boats could move along with mostly inexperienced young men paddling them.

We might have made it. However, just before we reached shore Plank jumped overboard. He was chest deep in the cold water loudly gasping for breath when the woods erupted with the crackle of rifle and machinegun fire. The muzzle flashes were so numerous that if live fire had been directed at us we would have been slaughtered before we even touched dry ground.

"Plank! What in the shit are you doing?" Our Squad Leader was standing up as he screamed over the thunderous noise.

"I couldn't sit there like a damn duck swimming along and waiting to get ambushed," Plank shouted over the barrage and resulting tumult of soldiers shouting, the occasional rifle fire from our side, and all the cursing and struggling of trainees trying to get out of the boats. Life preservers were frantically stripped off. Dozens of them lay floating at the water's edge, serving absolutely no purpose whatsoever other than to underscore the level of confusion which had taken place there only a few moments before.

When all the chaos had settled down, the referees graded our Company on how well we had conducted this part of training.

Company C received a mixed evaluation on the exercise. On one hand we were congratulated for making a quiet, professional river crossing at night, but the praise was obliterated by the conclusion that we would have suffered nearly one hundred percent casualties at the river's edge—thanks entirely to Pvt. Plank's ill-advised last-minute suicidal assault.

The culminating events of Basic Training were Field Bivouac and the Escape and Evade exercises held during that last week. The Company marched more than ten miles to reach the bivouac area, and then established a camp, which would serve as a base of operations for several days. Trainees set up two-man pup tents and then worked together to erect the much larger wall tents that would serve as the mess hall and company headquarters. It was January, unbelievably cold, and for some of the fellows from large cities their first experience at overnight

camping. What an introduction to the camping experience!

"I don't know if I can live through this," McCoy said as he and I struggled to drive the tent pegs into the frozen sand.

"Weren't you ever in the Scouts, or didn't your dad take you camping?"

"Manooch, you've got to be kidding! Remember, I came from the projects in Baltimore. Most kids there didn't have fathers who would take them camping, and there sure as hell wasn't a Boy Scout troop anywhere around there."

"Sorry. I guess I wasn't thinking. Here, let's get our tent ditched," I said as I grabbed my trusty entrenching tool as a cold rain began to come down in sheets.

"What's ditching?" McCoy said with a puzzled expression and rain pelting his face.

"That's when you dig a shallow trench around the tent so that it will catch the water and keep it from flooding inside and soaking everything. If you'd ever tried to go to sleep in a wet sleeping bag you'd never want to go through that again—if at all possible."

"Well, one thing's for certain," McCoy said. "We need to do something about all this water."

When we finished digging we stood up and looked around at all the pup tents that seemed to have sprouted like olive drab mushrooms from the ground. There had to be at least a hundred of them in order to accommodate a full infantry company.

"Come on. Let's go eat," Mc Coy said in response to the chow call. We got our "mess kits" out and headed up the hill. The kit is made of metal, and consists of two main parts like turtle shells—one on top and one on the bottom, secured in place by a folding handle. One component is usable as a small frying

pan with its handle; the other is the plate, partitioned down the middle with knife, fork, and spoon stored inside. The mess kit is carried in the soldier's backpack. The drinking cup is also made of metal, has a foldable handle, and is stored with the canteen. The canteen fits inside of the cup and both are worn on the right hip. This type of mess kit often clanked when we marched, and therefore had to rank right up there with white t-shirts and white name labels as being dangerously conspicuous when a soldier is trying to remain undetected.

We stood patiently in the rain in our dripping ponchos, waiting for our time to be served. You've got to give the U.S. Army credit for one thing. When the menu called for hot food it was *hot* food. It was good, it filled you up, and it made you warm inside. I'd bet there are few armies in the world that can deliver hot chow to the troops in the field. Of course, on occasion we were served K-rations or C-rations, but only during times when we were "blacked out," or in a stealth mode.

We remained in bivouac for a week, and the flu or some other malady swept through us like syphilis through Bangkok. There was an almost steady stream of troops going to and from sick call. That was simply part of it. When soldiers are in the field a percentage of them will be sick, injured, and even killed. In fact, in many instances as many men have been injured or killed by accidents as are casualties of actual combat. Thousands of aggressive young men, all of them doing dangerous things, can result in a high level of mishaps. Those in charge must plan for this in order to maintain troop levels that will enable the objectives to be achieved.

We might have been in a blacked out situation, but an enemy would have to be deaf not to hear all the coughing from hundreds of yards away.

"You know what, Manooch?" McCoy's muffled voice was barely audible through his sleeping bag.

"No. What?" I replied from deep inside mine.

"I don't have to worry about getting shot in some war

someplace. I'm going *to freeze to death* right here in South Carolina. I thought the South was supposed to be warm! I had a choice of going to Basic Training either at Fort Jackson or Fort Dix in New Jersey. I didn't want to go to Dix because I thought it'd be too damn cold."

"Well, you messed up! Now go back to sleep."

The Company was provided with several large latrines, all of them screened from view by large canvass curtains. The "pisser" was a terracotta pipe that was slightly angled and driven into the sand. It was designed for someone taller than me, and I had to position myself just right to gain access. That must have been a hell of a sight, seeing the shorter guys peeing on tiptoes. Pine straw was stuffed into the four or five inch opening so that when you peed you wouldn't splash yourself. Going to the latrine at night was not fun. There was no time to put on boots, and the only things we wore were our long johns and socks as we dashed for the latrine. It was too cold for snakes, but walking could be very painful because of the sand spurs common to that area. Even when we'd crawl back in bed, the small pricks made by the spurs itched like the devil.

The sand spur is one of the few things that God must've dreamed up when he had *absolutely* nothing else to do, and most other plants and animals had already been created—and for specific purposes. You could sit around all week and never dream up any good use for sand spurs. It's like some of those things that when you first meet God in heaven that you're bound to ask: Where did all my socks go? What are fleas, ticks, mosquitoes, and sand spurs good for? Why does something, like a pill, you simply drop on the floor roll off to gosh-knows where, never to be seen again? Sand spurs are in a special category by themselves, and although they may be found throughout the southeastern United States, South Carolina is undoubtedly the "sand spur Mecca."

About the same time that we were at Fort Jackson there was a high school in Wampee, South Carolina, not far from the North Carolina state line. The school is now the North Myrtle Beach Middle School—doesn't have the quite the same ring to it, does it? Back in the fifties and sixties, Wampee had a football team that was always out-manned in size and speed by their opponents. However God had graced the Wampee football field with the most wonderful crop of sand spurs you could imagine. Legend has it that the Wampee school said its football budget was so limited the team could only afford a few pairs of football shoes. The opposing team often felt sorry for the Wampee players so they would provide only several pairs of shoes for their players, leaving most of the players on both teams shoeless to make things more even.

The Wampee coach had his team run a half dozen or so plays right up the middle of the line for two or three yards, and sort of put the opposing team to sleep. When the time was right, Wampee's best running back and two blockers, all wearing football shoes, would sprint around the end right through a patch of sand spurs. The defense in its shoeless pursuit would be left limping far behind as the Wampee ball carrier crossed the goal line. This of course only worked one time, but that was often enough for a victory.

My mind shifted back to reality as I awoke after shivering through another freezing cold—practically sleepless—night to a dim gray morning without a promise of sunshine.

"Today's the day we do our survival and escape and evade training," Mahoney said as his smiling face peered through the opening in our tent. He seemed to always be up early and ready to go.

"How can you be so damn happy when you know we're going to be taken way back in 'God knows where' and dumped off with nothing to eat, and only a compass to get us back?" That sounds like pure shit to me." McCoy was talking—all the while on his knees trying to roll up his sleeping bag.

"Let's go get some hot chow," I said when McCoy had finished, and we grabbed our mess kits and hurried up the hill to catch up with Mahoney.

Following what was to be our last meal of the day, the entire Company was given instructions on how to survive alone if cut off from friendly forces, and how to avoid capture until reaching secure positions. To an infantryman, there's probably nothing more dreaded than being cut off from your own folks and then being surrounded and captured by the enemy. Many of the enemies that we face today are very cruel to captive soldiers. The brief introductions we received on this subject fell woefully short of providing enough details to allow us to accomplish the tasks of avoiding and escaping. And, we would probably have to survive without food.

It was too cold to find the types of wild foods that we were told were edible and could, if necessary, sustain an isolated soldier for several days. Lizards, snakes, ants and other insects were dormant this time of year, and we had little chance of catching what we considered to be more tasty critters, such as rabbits and squirrels. We were told that the tubers, or roots, of cattails could be eaten raw or boiled like potatoes. It was too damn cold for cattails to be emerging, and if we did find any we weren't about to wade through freezing water to dig up submerged roots.

We really didn't have enough time to even think about eating because we were dropped off, one at a time, over a distance of about two miles, and had "enemy forces" after us from the start. Most of the guys were captured in a relatively short period of time because all they did was run through the woods as they tried to get back to the designated pickup point.

I didn't get caught, mainly because I let most of the ruckus get way in front of me while I hid and stayed put for a couple of hours. I found a large oak tree that had blown down, exposing the root mass and a large partially hidden hole in the ground. I got in there. My hunting experience really paid off, because I could be very patient. When I hunted deer with a rifle I had to remain still for several hours at a time, and I also learned how to move quietly through the woods as I stalked the smaller upland game animals.

The closest I came to getting captured that day was when I had to cross a dirt road, which could not be avoided, to reach my destination. There were mounted patrols on jeeps that unpredictably traveled the road, and stationary sentries were positioned at two or three hundred-yard intervals. I crept to the edge of the woods and surveyed the road as far as I could see. Not fifty yards away was a creek and I crept up to it. It went through a metal culvert under the road and was just large enough to crawl through. I straddled the trickle of water by placing my knees and hands on either side. It was pitch black in there, but I could see light coming through the other end. Once clear of the road I carefully worked my way back to our encampment.

"Gung Ho was one of the first ones captured." Moody was beaming as he told how Plank at first tried to sneak through the woods, and then became impatient and grabbed a stick and charged the first enemy soldier he encountered. "They tied up his hands, put him in a truck, hauled him back here and placed him in that temporary stockade over there." Inside the wire barrier were fifty or sixty members of Company C milling around and obviously ready to get freed and fed. Plank was sitting off to one side by himself.

CHAPTER SIXTEEN

We had our graduation from Basic Training ceremonies at Hinton Field. Many of the soldiers' parents came. However, I honestly don't remember whether mine did or not. I do remember that there was a hell of a lot of us lined up on the parade grounds as we proudly "passed in review" and "eyes right" was the order when we went by the viewing stand. Major General Charles S. D'Orsa, Commanding General of Fort Jackson, and his immediate staff stood in recognition of the graduates' accomplishments. It was a good day. We were given free time to

do anything we wanted to do as long as we stayed on Base. We had already received our extended leave at Christmas, and were therefore not awarded the traditional after-Basic leave.

There were a number of stores (canteens) for on-base personnel that we could go to. These civilian-run oases sold things such as letter-writing materials, toiletry items, snacks, and—most importantly—beer and cigarettes. There were also "huts" on Base that sold hamburgers, fries, and soft drinks like fast food places do today. An E-1 private, the lowest pay grade possible, made eighty-one dollars a month. However, with food, shelter, and clothing provided, that was plenty—and there were not many places we could spend money. That afternoon several of us sat at a table drinking pitchers of cold beer and listening to recorded music.

"Tomorrow morning we find out where our next duty station will be."

"Where do you think you'll be assigned, Lazada?" I said as I poured myself another glass of dark beer. I didn't particularly like beer, but could get it down. In fact I won several bets by gulping down a glass in about three seconds.

"I'm staying right here at Jackson where I'll be entering Advanced Infantry Training. (*It's now often referred to as 'Advanced Individual Training.' but that's not what we called it.)* The fact that Lazada was going to AIT didn't surprise me because I could easily picture him as a full-time infantry soldier, maybe even going to airborne school at Fort Bragg or Fort Benning after that.

"Shit, that's more Basic Training if you ask me," Mumford chuckled. Today had been the first time in a long while that we had seen his face free of coal dust and sweat. He didn't look too bad all slicked up with a clean face and fresh shave.

"I'm going to Fort Gordon, Georgia to band school. I was in an administration and marching band company in Raleigh before I came to Jackson," I said, as proud as if I'd been accepted in medical school someplace.

"I wouldn't count on nothing until I heard it called out by the Sergeant," McCoy said as he pushed back from the table. "They'll be closing this place up shortly. Let's get on back to the barracks. The sooner I get my next ticket punched the better I'll feel."

Four of us were assigned to KP duty the next morning. We reported at 0330 and started helping the cooks get ready to prepare and serve the last breakfast we were to have as members of Company C.

The suspense caused by waiting for our orders was enough to make me crazy. The entire Company stood in formation as names were called out in alphabetical order, along with our next duty stations. Those of us on KP stood next to open windows in the mess hall so we could hear the final roll call. Some of the fellows were going over to another Regiment at Jackson for OJT (On Job Training), which included things like clerk-typist school. Others were going off to places like Fort Polk, Louisiana for more infantry training or cooks' school, to Fort Gordon, Georgia for military police or band assignments, to Fort Sill, Oklahoma for artillery training or to Fort Knox Kentucky for tank school. When the Sergeant finally got to "M" I could hardly stand it. For some reason I had a very bad feeling.

"Manooch, Charles, the third. NG24970479. Fort Jackson, Company B, Twelfth Battalion, Third Training Regiment."

It took a few moments before what had been called out recorded in my brain. And then it hit home hard. *AIT? AIT!! There has got to be some mistake. I can't be going to Advanced Infantry Training for four more MONTHS of this punishment, can I?* That day was one of the longest I endured at Jackson. I couldn't wait to get off duty that night and call Dad to see what he could find out. There had to be a mistake and I was sure he could handle it for me. I had to tell myself to calm down and that I'd be going to Fort Gordon after all.

That evening after I'd been released from KP, I ran over to the nearest pay telephone. "Hello, Dad? I'm calling from a pay phone down here so I don't have long to talk. Today they called out my name as being assigned to AIT. Can you believe that? Can you do something to straighten this mess out?" I said in what had to be a pleading voice.

"Nope. Your orders have been changed and AIT is where you've been assigned."

"How in the heck did that happen? I'm supposed to get BAND training, for God's sake!" I was already looking forward to going to Fort Gordon!"

"I had your orders changed, son. Remember when you enlisted I told you you'd either love the military and would go on to Officer's Candidate School, or you'd choose another profession and want to go back to college? Remember that? Advanced Infantry Training and then Basic Unit Training is the best course of action for you to be considered for officer training."

I could hear Mother in the background hollering. "Charlie what have you done? What's happening with Chuck?"

"The best damn thing that could happen that's what. This may be his last chance to become a man and straighten up. He's up to his neck in it now, and he'll either sink or swim."

"Okay, soldier. That's long enough. Get off the phone!" There were five or six guys standing in line waiting to use the phone, and right behind me was a big Sergeant. I hung up and dejectedly walked back to the barracks to start packing. I had no other choice. The road ahead of me had been clearly marked and I was in no way pleased with the signs.

CHAPTER SEVENTEEN

"Jesus Christ! Can you believe that?" I looked up the street where Johnson was pointing. There were several soldiers from the Fourth Regiment standing on the corner apparently waiting for a bus. They wore dress greens and it was sadly obvious to us that they were leaving on three-day passes. We sure as hell weren't. It was Friday, mid-afternoon, and we were standing in formation in front of our new barracks waiting for our Sergeant to proclaim his authority over our lives for the next two months.

"I'm Sergeant Talbot, and I'll be in charge of this platoon as long as you are here. You'll do *what* I say, *when* I say. You got that?"

"Yes Sergeant," we all responded. Johnson, Mahoney, McCoy, Mills, Toby Lee, Moody, Lazada, and about thirty other guys from our old Basic unit had been assigned to Company B, Twelfth Battalion, Third Training Regiment for Advanced Infantry Training. The Regiment was made up of four battalions—the Eleventh, Twelfth, Thirteenth, and Fourteenth.

The Sergeant was obviously tough as hell—we could clearly see that. He *had* to be because of his size. He stood about five feet four, and was wiry, weighing somewhere around one hundred and twenty pounds. He looked like a weathered short cowboy. The strangest thing about him though was he had a small monkey sitting on his shoulder. "This monkey's name is 'In-Action.' and I got him in Guam when I was returning from Korea. When I yell 'in action' you-all, and my monkey, will stand straight up—at attention. When I yell 'outta action' all of you, and my monkey, will drop to the ground. Sometimes we'll go through this several times in a row and I promise you that you'll get tired long before my monkey does. The little bastard can jump around and climb up and down things all day."

Sergeant Talbot proceeded to introduce us to our next phases of training. "The Third Regiment will be your home for the next eight weeks, give or take a few days. When you complete AIT you'll move on to Basic Unit Training—also here at Jackson. Both programs are designed to train you as a MOS 111, Light Weapons Infantry Leader, or a 112, Heavy Weapons Infantry Leader. Since you are in my platoon you're obviously a light weapons soldier. Some of you are Active Army, some

Army Reserves, and some National Guard. Regardless of which group you represent, each one of you will receive the same level of training—it won't be easy. If you finish my training, you will be a top notch U.S. Army Infantryman."

When dismissed we were told to put our gear in the barracks, get out our steel "pots" (combat helmets) and then regroup in the Company street. I'll never forget what we were ordered to do next. We had to fill our helmets with sand, and there was no scarcity of it, and go throughout the barracks pouring it on the floors. The floors were *immaculate before we started.* They had been waxed and polished to a dark green sheen by the former inhabitants. Once we covered the floors with sand we used heavy handheld brushes and waxing machines with brushes to remove all the fresh wax from the floors. You guessed it. Then, we had to rewax and polish the floors. The barracks had also been freshly painted, and we were ordered to use razor blades to scrap away paint that had been left on the window glass. These discipline chores took us to supper time before they were completed.

"It's hard to believe that here we are working like dogs and most of the fellows assigned to other units are happily off on three-day passes." Toby Lee and I were standing in line outside the mess hall, and we could smell the hamburger steak and onions being served up ahead. Right out of the blue, Toby Lee said, "We're getting new rifles." What brought this up no one knew.

"They're the newer M-14s. They can fire on fully automatic as well as semi, and the bullets are those 7.62-mm ones used by all N.A.T.O forces. We're also getting other, newer types of weapons that will be issued to each squad." Mahoney overheard what Lee said, and felt compelled to offer his "two cents worth." Barry was smart—there was no doubt about that, and as usual seemed to know what and when something was going to happen—and long before the rest of us did.

We soon learned that several things were very different here in our new Company compared with Basic. For the first time most of my closest friends and I were assigned bunks in the lower bay in the barracks. Also, like Sergeant Talbot had told us, there were three light weapons platoons and one heavy weapons platoon. We were in one of the light weapons platoons and each squad in the platoon was issued an M-60 machine gun, which fired 7.62-mm rounds, and an M-79 grenade launcher. This thing loaded like a single shot, open breech shotgun, but it could accurately fire an exploding shell through a window at about two hundred yards. Up to this point we had been trained using the old inaccurate rifle-mounted grenades that were attached to the muzzle of the M-1 rifle. They were usually fired by putting the stock of the rifle on the ground or by positioning it on your thigh. The NCOs warned us that a soldier's femur could be broken if the rifle stock was not placed and held just right. We had practiced firing these grenades during Basic Training at clusters of silhouettes of simulated soldiers. After firing the grenades we inspected the targets to note the extent of the damage, which extended out to a radius of about fifteen yards.

Guys in the heavy weapons platoon carried M-14 rifles like the rest of us, but they were also provided with two types of mortars—60-mm and 81-mm—and 106-mm recoilless rifles mounted on jeeps. These troops were the "heavies" in the Company, and together with our three light weapons platoons, created a fully equipped fighting unit.

"You know what, Manooch? Remember when you, Mahoney, and I were work-detailed to that 'Infantry—Queen of Battle' range? Well, we're beginning to look like those fellows that showed us all the firepower." McCoy was right. We had the look, down to the brown, blotched patterned camouflage that they had worn on their helmets that afternoon. Whether we liked it or not, Sergeant Talbot and his monkey were committed to the task of assisting us become well-trained infantrymen.

CHAPTER EIGHTEEN

Next morning we really got into the swing of things. Sergeant Talbot got us up, stayed in the barracks long enough to casually check our footlockers and bunks, and then turned In Action loose. That little snitch ran around, jumped up and down on our bunks, and I swear—pointed out things that weren't just right. This made me think of what it must be like hiding in a spaceship with a little evil alien stalking around looking for you. When the monkey and Talbot saw our imperfections we all had to pay the price. The Sergeant would yell, "Outta action!" and we fell on the floor. We stayed there a few seconds and he'd yell "In action," and we jumped back up. This little exercise became a daily ritual, and not surprisingly, it wasn't long before we hated that monkey's guts. Guys in the other platoons started referring to us as the "monkey troops" of Company B. You could bet they sure were glad as hell that they didn't have a monkey-owning platoon Sergeant.

Before we could begin our training in earnest, something important was taking place far away that would forever make Fort Jackson a part of U.S. history. A year earlier the American military, and in particular the Central Intelligence Agency (CIA), had assisted Cuban exiles in an attempt to overthrow the Castro-led Cuban government. In April 1961 the Bay of Pigs (*Playa Giron*) Invasion was launched on the beaches of southwest Cuba. Participating Free Cuban troops were part of the 2506th Brigade, which was formed by soldiers previously exiled from Cuba. The troops had trained under U.S. military supervision for almost a year at locations throughout southern Florida and in Guatemala.

The landing was a fiasco "from the get go" for those actively participating, and was also a major embarrassment for the government of the United States. There were about 1,500 men in the 2506th trying to defeat perhaps 50,000 Cuban troops loyal to Castro. The plan was to *achieve immediate superiority in the air* by eliminating the Cuban Air Force on the ground, and then gaining popular support from the people to overthrow the Communist Government. Unfortunately for the 2506th, the bombing of Cuban military airbases and strafing of troops on the ground was unexpectedly terminated by orders from Washington.

In addition, Castro had prior knowledge of the invasion, and had moved many of his aircraft to other locations. He was also able to utilize the expertise of Russian military advisors to organize and direct a Cuban infantry counterattack, aided by tanks. The 2506th didn't have a chance without support from the air, leaving the invaders trapped on the beach and in nearby swamps where they were easily captured. Approximately 1,200 were captured, quickly placed on trial and convicted. A few were executed, including some American pilots, and the rest were sentenced to thirty years in prison.

In May of 1961 Castro offered an exchange of prisoners to the United States for five hundred bulldozers, food, and medicine. On December 21, 1962 the exchange was finally made, and President Kennedy met the returning brigade at Palm Beach, Florida on December 29th. The remaining members of the 2506th were posted at Fort Knox, Kentucky in early 1963—when we were in AIT. They didn't stay in Kentucky long. Fort Jackson became their home, and they were the subject of many conversations.

"I heard those losers were refusing to come out of their Kentucky barracks because it was too cold and that's why they were sent to Jackson's Third Regiment. It *sure* as hell's not too warm *here* either right now," Johnson was talking to several of us as we stood in line for some reason or another. We were always *standing in line.* That's part of the soldier's saying that the army routine is "hurry up and wait."

"They might have been 'losers' alright, but it was *our government* that got them all fired up, trained and ready to go, and then left them stranded down there without a chance in hell.

Besides most of them have *already been in combat*." Mahoney was right and we all knew it. We were willing to give the Cubans a chance. We'd wait and see.

CHAPTER NINETEEN

One of the first elements of our new training schedule was an introduction to light infantry weapons that had been used in the Second World War and the Korean Conflict. Some were still in use. Unlike in Basic Training, we were allowed to travel by trucks to participate in most field exercises because Lord only knows we'd already proven that we could march "over hell and half of Georgia." We arrived at a large parking and unloading area surrounded by firing ranges, some short and some long. We jumped out of the trucks and were organized by platoon. It was obvious that we would be training in four separate groups.

"Okay, guys. Listen up. We are going to be 'live firing' today, so remember to practice the safety that you've learned up to this point in your training. Some of the weapons that you'll be using can cause big time damage, and a misstep could result in injury or death to those other than yourselves." The Sergeant had obviously repeated those same words a thousand times. As he was instructing us we craned our necks to grab glimpses of the different firing stations located all around us.

"Damn, Manooch. Look over there. What do you call those things? I've seen them in the picture shows and they're bad ass." I looked to where McCoy was pointing and clearly saw what he was referring to.

"Those are flamethrowers. Good Lord! I'd hate to be in front of one of them with somebody mad at me."

"Do you think we're actually going to fire those things?" McCoy's eyes were as big as golf balls.

"I guess so, or they wouldn't have them lined up over there," I said as I saw Sergeant Talbot glaring at us. "We'd better shut up, or you and I will be used as targets."

The first weapon we fired was the .30 caliber M-1 carbine. "These are small, lightweight weapons used primarily by officers in World War II and in Korea. They are semi-automatic, clip-fed, and are good for in-close fighting. The rounds are short, and the brass looks like an oversized .22 caliber shell." The Sergeant in charge of the range was giving us a description of the short rifle he was holding. We'd seen them in war movies and in newsreels. John Wayne had held many a one of them.

"The major drawback to this weapon is its lack of knockdown power. Soldiers returning from Korea complained that enemy troops could often continue to fight even after being hit by carbine fire. The positive points are: The carbine is light, thus easily carried, has a relatively rapid rate of fire, and is more accurate at a distance than a pistol."

We fired the carbines, using the prone position, at targets that were placed at about two hundred yards. The rifles were light as described, and had very little recoil. Hitting the targets was not too difficult.

Next on the list was the .30-06 caliber BAR (for Browning automatic rifle). I have never fired a weapon before or since that I like as much as the military BAR. The rifle is fully automatic, is clip fed with the same ammunition used in the M-1 rifle, and comes equipped with a folding bipod, which can quickly be put in place when the shooter is in the prone position. Historically this weapon, like the M-1, has been one of the most well respected rifles in the U.S. Army arsenal. It was used in Korea and in both the European and Pacific theaters during World War II.

"Most of you will find that the best way to fire this weapon is to sight the target, start firing, and observe the placement of the rounds as they strike the ground around the target. Then 'march' the bullets up or sideways until they cover the target." The Sergeant in charge of this range was pointing at white targets that were placed in front of the firing positions at a distance of about four hundred yards.

"Hell, I can hit the damn target with the first shot," Lazada snorted as he got down on the ground beside his BAR. He couldn't. But like the rest of us, he was able to follow the Sergeant's guidance and moved his bullet strikes until they covered the target.

I loved the BAR. It chugged along like a slow-firing machinegun, but once on target it could really cause some damage. The major drawback was that it was heavy, and I could only imagine what it would be like in combat if it had to be carried while running in a hot jungle or up a steep incline. However, once brought into action, the rifle could really assist an infantry assault, or defend a position.

I didn't look forward to our next firing exercise, and as we marched over to the range I remembered how Dad used to let me fire his service pistol. He was an officer and therefore was issued a semi-automatic pistol. He let me shoot at cans and bottles, but I couldn't hit the broad side of a barn. I don't know why, because even as a small boy I was a good marksman with a rifle and shotgun. When we arrived at the range we were amazed at how close the targets were in regard to where we would be firing. The Sergeant in charge of the range was standing there ready to receive us.

"You guys will probably not fire a pistol as a soldier after you leave this range. However, we want you to at least have some familiarity with the weapon so that if you have to pick one up and fire it in a combat situation, you will at least know how the damn thing works." He held out one of the automatic pistols so we could see it.

"Officers carry these in combat. Or, they're *supposed to,* that is. Some take advantage of the situation and resort back to their childhood fantasies by carrying pistols that look like those worn by cowboys. You remember General George Patton, don't you? Hell, during the Second World War he carried nickel-coated revolvers that had ivory handles."

"This morning you'll be shooting the .45 caliber, Model 1911 Colt semi-automatic pistol. It is fed by inserting a clip up into the handle like this." He demonstrated loading and cocking the pistol and then continued: "It has been used by the Army for a long time." The Sergeant had our attention as he stood by a table where about twenty pistols were lined up on top. They *looked* like military pistols. They were black and appeared to be very sturdy.

"Let me tell you a true story about this weapon. Back during World War I this pistol gained its proud reputation when it was used one October day to fight the Germans. A U.S. Army unit was pinned down by machinegun fire. I think it was in France someplace. Anyway, key NCOs were either killed or wounded to the point that a Corporal Alvin York, a young man from the mountains of Tennessee, assumed command. Single handedly he worked his way up a hill until he had flanked the German position. With the aid of a 1917 Enfield bolt-action .30-06 rifle, and a Colt .45 he killed twenty-five Germans and captured another one hundred and thirty-two. He received the Medal of Honor as well as a number of other decorations."

Six or seven of us at a time took positions at shooting stations, and were given individual instructions on how to load, operate the safety, and fire the pistols. One of the amazing things about this particular range was that the distance to the targets was measured in inches instead of in yards. Converted, the distance seemed to be about fifty feet. I still couldn't hit the target with the pistol, although the Lord knows I tried hard enough. And the damn thing kicked like the devil. It was odd, but some of the fellows who were not expert with the M-1 were excellent shots with the .45 Colt pistol.

Right after firing on the pistol range, we marched over to where chow was being served. Our Company mess hall cadre and KPs brought hot lunch out to the field. We were served from insulated containers that were lined up buffet style and we ate from our mess kits. The old saying that "the infantry marches on its stomach" is undoubtedly true, and knowing that *hot* food will be served is very good for morale.

As usual, the food was good, and after we had eaten we lined up to clean our mess kits. The cup, knife, fork, and spoon have slots in their handles, and the plate has a small metal loop, which allows all of them to be stacked on the small frying pan handle. The entire kit can be dipped into boiling water that contains detergent. The washing and rinsing stations were in separate metal garbage cans that had large kerosene heaters submerged in them. The portion under water looked like a giant metal doughnut, equipped with a smoke stack attachment that stuck up over the can to a height of about seven feet.

"Come on fellows. Hurry up! Form up over here on me. We're going to introduce you to a couple of big boys now." Sergeant Talbot and his monkey were standing next to two other NCOs, and as usual he and his monkey had big toothy grins on their faces.

"Damn. I recon we're going over there to shoot one of them flame things."

"McCoy. Sometimes you amaze me with your brilliance," Johnson murmured as we did a right face and started marching to the next range, which was within sight. As we got closer we saw a Sergeant standing next to a three-quarter-ton truck with its tailgate down, and three flamethrowers lined up as if ready to take off. We were halted in front of the Sergeant and executed a left face so he could instruct our platoon.

"This weapon was used extensively during the Second World War. The Germans first introduced one like it to destroy Dutch gun emplacements that were difficult to assault otherwise, and we then developed one of our own. It's called a M2A1-7, man-pack flamethrower, and you can imagine what fear it can cause to enemy infantry. You've all seen the newsreels in picture shows where U.S. Marines brought gruesome destruction to Japanese soldiers hiding in caves on Pacific islands. It *fried* them, and Marines said you could hear the Japs sizzle and pop." This last bit of information was more than we really needed at the time, but it did get our attention.

The Sergeant continued. "The flamethrower is made of two major parts—the backpack, which contains two metal cylinders holding compressed gas, like nitrogen, and flammable liquid; and the gun, or nozzle, which is a spring-loaded valve with a trigger. When you press that trigger—*WHOOOSH.*" We all stood there with our mouths open as the NCO gave us a live demonstration—and we felt the heat.

"Heaven help us. There's no limit to the extent that men will go to think up ways to kill each other." Mahoney said aloud what most of us were thinking. I had a hard time imagining a worse way to die than to be burned alive. Hell, this burning business goes back to when the Christians were killing Jews in the 1300's because they wouldn't accept the Holy Trinity—and I guess *long*, *long* before that. You'd think by now we would've developed higher moral and ethical standards, but obviously we haven't. *To be perfectly honest, I no more stood there philosophizing about human values than I flew off to the moon. I wanted to put the darn thing on, flame the target without burning myself up, and then take the thing off. The rationalization part didn't come until years later.*

Each of us took our turn struggling to put the backpack on, and then actually firing the flamethrower. The backpack was heavy and we had to back up to the tailgate of the truck, squat slightly, place the two straps over our shoulders, then straighten up and slowly walk away from the tailgate. It was then that the realization hit home that gallons of a flammable liquid were on your back, and you were about ready to turn on an igniter. Common sense tells you that this isn't smart at all. However, I must confess that when the fifty feet or so of fire shot out from the nozzle, the shooter feels not only the heat, but also the power.

The last weapon we fired that day was what is commonly called the "bazooka." Like the flamethrower, it was used extensively during the Second World War and in Korea. "This weapon has two bad ends," the Sergeant was telling us.

"Naturally, the front end is the business end and that's where the rocket comes out. But the rear end has a tremendous back-blast that can kill or seriously injure someone standing too close. You have to look both ways before you fire this thing. The proper name for it is the 'M-9 two-man portable anti-tank rocket launcher.' Let me tell you about it. The weapon is essentially an open-ended tube that's about fifty-four inches long and weighs about fourteen pounds." We could tell that he'd given this little lecture dozens of times, and had the statistics down pat, as he continued. "It is typically used against enemy tanks, and has a range anywhere from fifty to seven hundred yards. However, it's difficult to imagine trying to hit a target at seven hundred yards. Firing it is a two-man operation. Two of you come up here to demonstrate. Manook and Moody. You-all will do."

The Sergeant had me kneel down on my right knee and he placed the rocket launcher on my right shoulder. He took one of my hands and had me grip one handle, and placed the other hand on the trigger handle.

"Now Moody you get down behind Manook's right shoulder where you can insert the 60.07 mm rocket into the tube. Be careful that you are completely out of the way—off to the side so that when it's fired you don't catch any of the back blast."

Each crew took turns firing dud (non-explosive) rockets at a stationary target placed at a distance of about one hundred yards. I don't remember if any of us hit it. The Sergeant placed a wooden crate behind Moody and me to demonstrate the power of the back blast. The crate was destroyed.

Once every soldier in the Company had fired each of the different weapons, we were transported back to the barracks. We knew that our coordinated unit training would begin the next day. That's when, as they say, "the rubber meets the road." Advanced Infantry Training and Basic Unit Training, unlike Basic Training, strives to incorporate the individual soldier into thoroughly trained squad, platoon, and company-sized fighting infantry units. The individual becomes a team player.

CHAPTER TWENTY

While we were beginning our AIT cycle, the Cuban Training Program was being established on Base in the 1st Training Regiment. Most of the Spanish-speaking cadre from Fort Knox had arrived at Fort Jackson by early January 1963, followed a few days later by approximately 1,700 Cuban trainees. These guys had been in various stages of their training program at Knox that was supposed to last twenty-two weeks. We were told that administrative mix-ups had resulted in the transfer, but we also heard that the transfer was made after the Cubans refused to "fall out" of their barracks for training. This less than flattering explanation seemed to be substantiated during the next couple of months. Initially, Cuban volunteer input to the Program was good, and in early April there were fourteen companies totaling more than 2,700 soldiers. However, by the end of July there were only enough Cuban troops to fill one company, and the program was terminated.

There were other problems as well. "Those guys sure get sick a lot," Mahoney informed several of us one afternoon while we were policing the Company area. "I know a fellow that works as an orderly in the Base Hospital and he tells me that the wards are overflowing with what are supposed to be 'sick' Cubans. He says they don't look or act too *sick* to him."

"They're in there to get out of work just like they did at Fort Knox."

"You're at least *partly* right Johnson," McCoy said in a sarcastic tone. "But I'll bet the *real* reason they're in the hospital is to look at those cute nurses' butts in those tight uniforms of theirs. You can bet there's a whole lot of 'jacking off' when the lights go off at night." We could all relate to that. Masturbation, in general, is something you'd certainly expect from a bunch of eighteen and nineteen-year old boys corralled off somewhere without any female company. It's only natural. However, masturbation in bunk beds is extremely difficult because your bunkmate, upper or lower, would certainly be alerted to the action because the beds would rattle and shake. And, the Lord only knows that no one wants to be kidded in public for "beating his meat." An NCO with that bit of personal information could make life miserable for some poor fellow.

For whatever the real reason, the Cuban volunteers were

indeed admitted to the hospital at a much high rate—four to five times higher compared to their American counterparts. Various reasons were proposed for this disparity, such as less resistance to diseases, malnutrition prior to arrival at Jackson, and being unaccustomed to cold climates. The fellows in my platoon still supported the slackness and lust arguments. I was soon going to get a better look at this for myself.

A "live fire" exercise involving troop movements is something we had heard about, but had not taken part in yet. That was about to change. It was late winter, so we had on our brown-splotched helmet covers, which supposedly helped us blend in with the drab winter landscape. There was no doubt that the sparsely forested land on the Fort Jackson Military Reservation was "drab"—regardless of the season.

The unit live fire training course involved squad-size units firing while moving up a hill to capture an enemy position. This would be repeated several times that day. We had completed

the first maneuver in which our squad had slowly advanced up a slight hill covered with shoulder-high turkey oaks. All of us carried M-14 rifles, which we fired on the semi-automatic setting. It was imperative that we maintained contact with those on either side, stayed in a horizontal line, and moved forward at the same rate. McCoy was about ten yards to my right. Mahoney was about the same distance to my left. It was cold as usual and all three of us were "blowing smoke" as we cautiously moved through the scrub oaks.

"Damn it, Manooch. Slow the hell down. You want to be shot in the back?" A Sergeant was assigned to each squad and was responsible for squad safety. "Don't look back at me," he ordered. "There's nothing back here. Look straight ahead and keep firing when you have a target." Every now and then a target shaped like a human silhouette popped up and was immediately struck with numerous 7.62-mm rounds fired from eight or nine rifles. Plank was located about two guys over to my left, and he was so happy he was almost skipping up the hill. "Hell, this is what it's all about," he said as he opened up with a rapid, unrestrained burst.

"Get that damn thing off automatic!" the Sergeant screamed as he ran over to Plank. "One more stunt like that and I'll have your ass, boy." He meant it too. Our attention was suddenly diverted when the loud rapid chatter of a machinegun cut loose just up ahead. We dropped to the ground and lay there.

"Hope to hell that thing's firing blanks," Mahoney said with his face pressed so close to the ground that we could barely understand him.

"It has to be," I replied, "or most of us would be *dead.* That gun's only fifty yards up there."

In actual combat, the squad leader would have probably sent two guys crawling ahead, each carrying hand grenades and approaching the gun position from different sides. However, we couldn't do that today because the flanking soldiers could have shot each other. So we low-crawled towards the machinegun, firing as we advanced. We "captured" the gun, which we discovered was behind an earth (actually sand) and log bunker.

Later that day we prepared to make an assault on another fortified position. The entire Company was gathered in one of the staging areas, and I was sitting on a log picking dried weeds to stick into the band on my helmet cover. As I grabbed at the grass, a snake, which had been trying to warm itself in the sun, struck my hand between the base of my third and fourth fingers. I jerked my hand back, shaking off the snake, and saw the "U-shaped" bite mark. Ferguson was sitting beside me and he started yelling, "Manooch's been bit by a snake!"

The bite was bleeding pretty badly, but by wiping away the blood I could tell that the bite was made by a nonpoisonous species. I had been around all kinds of snakes when I was a kid, and there was no question that this was not life- threatening. Unfortunately, the Range Officer over-reacted. Total mayhem was the result, and I remember thinking that if a black racer snake could cause this much panic and confusion in the ranks what in the hell would happen if we were confronted with a real threat.

Since this was a live-fire training exercise there was a tactical ambulance right there on site, and before I could dissuade the medics I was plopped on a stretcher, loaded in the ambulance, and sped off to the Fort Jackson Hospital. There, the ambulance came to a screeching halt, and the rear doors were thrown open. You would've thought there had been an assassination attempt on the President of the United States. Several doctors and a handful of nurses were eagerly awaiting my arrival.

"This is the first snake bite this year in the entire 3rd Army," one of the doctors proudly explained. "We don't even have any anti-venom on hand. Nurse Pascal, call CDC in Atlanta and tell them to air ship some up here ASAP!"

"Yes, doctor." And with that she ran back inside the emergency room.

Anti-venom! Hell they're crazy as loons. They could give me an injection of Kool-Aid—it would do just as much good.

The gurney, with me aboard, was making fast time through the ER doors, down a short hallway and into a curtained-off cubicle.

"Move quickly, people. These things can be deadly in a matter of minutes—depends on the type of snake and physiology of the patient."

"Doctor, I was a bitten by a black racer, and it's not poisonous."

"Don't try to speak, son. It'll only make matters worse. Just try to relax. Where the hell are the monitors? I want BP, body temperature, EKG—the works. And, let's get that anti-venom here ASAP. This could be life or death."

"Doctor all his vitals look excellent. His blood pressure is up some, but that's probably from all the excitement."

"Okay, okay, nurse. Put him in Ward C, and let me know as soon as the anti-venom arrives. We'll give him that just to be on the safe side." The initial wave of anxiety was beginning to subside.

I was wheeled into the ward, still on the gurney, and wasn't too surprised to see that Ward C was cram-packed with Cubans. However, you would not have known it by looking at them. Most Spanish-speaking people I'd seen up until that moment had been either Puerto Ricans or Mexicans and they were usually dark—dark eyes, dark hair, and dark skin. They were typically small in stature, and to me they just looked foreign. Cubans didn't. They looked like you and me. Some black but most were white. There were red headed ones, blond headed ones, and some with dark hair—just like typical Americans. Some even had freckles and looked like they were from Mayberry. However, very few of them knew enough English to carry on a conversation.

The anti-venom was delivered to the hospital within several hours, and I was given an injection in my rear end. They could've just as well have shot it into my mattress for all the good it did. The next day my dad arrived, completely unannounced to me, and was accompanied by a doctor, nurse, and Sergeant McGee, a photographer for the 30th Division, North Carolina National Guard.

"Major, your son's injury represents the first snake bite of the year for the entire Third Army. Could possibly be the only snake bite for the year for all we know." The doctor seemed happy that this was the case, and also the fact that I was out of danger and would surely survive the encounter.

"How do you feel, son? You surely look fit enough," Dad said as he shook my hand and Sergeant McGhee popped off a number of flashbulbs. The next month edition of the *Tar Heel Guardsman* magazine had a picture of me sitting up in bed with a U.S. Army hospital blue robe on and Dad shaking my hand.

The brief caption, which appeared beneath the photograph, explained the serious nature of our meeting.

I was returned to barracks and was met by several of my buddies. "Good Lord, Manooch. When you get out of training for a couple of days you sure do it first class." However, Johnson and the other fellows seemed genuinely glad to welcome me back to AIT You know what they say. "Misery loves company."

CHAPTER TWENTY-ONE

For some reason that I'll never be able to explain, I was assigned to be the machine gunner for our squad, and was issued an M-60. There I was, at five-feet-seven, lugging the M-60 and Mahoney, at six-feet-five, carrying two ammo boxes. The M-60 is belt fed, so the ammo cans had 7.62 rounds inserted into what looked like metal ribbons folded up inside. When I opened the gun's breech, Mahoney fed in the ammo belt. This was to be done in rapid fashion so we practiced this procedure over and over. On maneuvers, we typically moved down dirt roads in two parallel files, and when ordered to form a defensive perimeter we'd run into the woods on either side and assume prone positions. Each member of the squad had a particular place he was supposed to be, and the machine gun was usually positioned on the left flank.

We were getting way beyond individual training, and were now concentrating on unit tactics. Each squad—as best as I can recall—had a Squad Leader, a radioman, a guy equipped with an M-79 grenade launcher, a machine gunner and his ammo carrier, and several riflemen. When the squad was on patrol, one rifleman was ordered to serve as "point." This soldier had the misfortune of walking out in front of the rest of his squad—thirty yards or so ahead. The distance was dictated by how open the terrain was. In combat, the point man was the first to encounter an ambush or mine. Life expectancy for this individual could be short.

The weather was getting warmer, so training outside was not as miserable as it had been just a few weeks earlier. Squeezed in between our unit training exercises, we received passes that allowed us to leave the Base for two and sometimes three days. We looked forward to these, but they were usually of such short duration that it didn't allow us to visit home—unless home was less than one hundred miles away.

None of us had been out of uniform for weeks, except for showering and sleeping. That gets old. Even when we went to church, or to the little "huts" scattered around the Base that sold hamburgers and hotdogs we wore fatigues and boots. To make matters worse, we'd very seldom laid eyes on a woman, except at the hospital, or off Base when we were assigned to some detail that carried us by a farmhouse. In these instances we'd strain to get a look—and not a single woman or girl looked unattractive to us in the desperate condition we were in. Thoughts centered around girls and sex constantly. I guess one way to describe Company B at this time was four barracks packed with testosterone and clothed in olive drab.

"You know what we ought to do in Columbia?" Mahoney said one day as we stood in line outside the Orderly Room waiting to pick up our passes.

"No. What?" I replied knowing there wasn't much we really could do in Columbia—our only means of transportation being buses.

"I think we should get a room at the DeSoto and 'entertain' a couple of 'ladies.' That sure would be a pleasant change of pace."

"I'll go into town with you, even get a room at the DeSoto, but the only thing I want to *get* is a good night's sleep in a big bed, and sleep as long as I want to without somebody yelling in my ear." However, while we were in the Orderly Room we both picked up "rubbers" that were made available to soldiers going on leave. Although I wasn't planning on using one, I didn't want the Sergeant to know that and let the word out. Earlier in our training we had received a lecture on the health consequences associated with venereal diseases. A series of pictures depicting infected male genitals had made this part of our training *unforgettable.*

The old DeSoto Hotel, located in downtown Columbia, had the reputation as being a relatively dignified looking whorehouse. Many years before that, however, it had served as a very respectable establishment where travelers could stay and dine and ballroom dance to live music, mingling with local residents who came there on special occasions. While I was at Fort Jackson many of the hotel patrons were certainly not "on the prowl" for female companionship. Some, particularly military personnel, sure were.

Passes in hand, Mahoney and I boarded a bus that would deliver us to our downtown destination for fun and relaxation. We wore the dressy Class A uniform as we stepped off the bus. "The first thing I'm going to do is get me a pair of pajamas. I haven't had any on since our Christmas leave," I said as I went into a large department store which obviously carried men's clothing. Once inside, I made my selection—nothing fancy, but comfortable looking. I paid for them and stuffed them into my small grip that I had purchased on Base. Mahoney was already outside and walking up the street.

The hotel was an old brick building with about eight floors, and like many old hotels it stood on a corner.

"You guys are fortunate that we do have one room available and it has a double bed. We could move a temporary roll away bed in there if you'd like." The desk clerk was being nice, but it was just a busy time and the hotel was booked to the point that if we stayed there we'd have to share a room and bed. Under the circumstances, that might prove to be very interesting.

"Both of us staying in the same room and you wanting to sleep, and I'm looking for some 'action' should be exciting," Mahoney said as he signed the register for one night, one room, and one double bed.

Action is what he got. Two young women just happened to drop by the room at about eight o'clock that evening, and they couldn't believe that only one of us wanted a "date." Mahoney and his friend played around in the bed while I sat out on the fire escape with my new pajamas on and watched the pedestrian and vehicular traffic several stories below. It didn't feel too uncomfortable out there. The weather was not cold or hot. The smells and noises of the city rose up to me from below, and I was intrigued by that for a while.

"Mahoney, please let me know when it's okay to come back inside. Hell, I paid for half of this damn room you know." I was afraid that the two of them would go to sleep and then it would be much more difficult to get his "guest" to leave. Finally I heard sounds coming from the bathroom and I hoped this was the early indication of departure. Thank goodness it was.

"Okay Manooch. It's all clear. " Mahoney called out in a most pleasant, relaxed, and pampered-to, tone. I crawled through the open window and saw him propped up in bed on two pillows. The only thing that I could see him wearing was a big smile.

"Give me one of those damn pillows and then get up and at least put your shorts on. We might have to share the same bed, but I'll be damn if I'm going to sleep with you like that."

We slept as far apart as possible—me clinging to one side of the mattress and he to the other. And we slept late. We went downstairs for breakfast, which the hotel served buffet style. For the first time in weeks we ate all that we wanted and took our time doing it. We paid our bill and then spent most of the day window-shopping for things we weren't allowed to keep on Base, or couldn't afford in the first place.

"What are you going to do when you get back home?" I asked seriously as we looked out of the bus windows at houses we passed by.

"Probably try to finish my college education. You know I have one year behind me and my grades were good enough to continue. I was just *tired of school*—grade school, then into junior high, then high school, and right off to college. That's a hell of a lot of time in the classroom. But I think I'm ready now to go back."

"I never even gave school a chance," I said as I turned and looked directly at him. "I was a goof-off the whole time—right through my first year at college. When I enrolled at Louisburg Junior College it was like I had never been to high school. It was a miracle, and the lobbying of my mother, that I got accepted in the first place. Honest to God, the only thing I remember academically from high school was sitting in civics class and discussing the meaning of a word. My teacher at Broughton for that class was Miss Runnion. She asked me to say the word 'prejudice,' which was printed in our texts, and tell the other students what it meant. I said 'pre-justice' and everyone starting laughing. Miss Runnion came to my rescue."

"He's mainly correct," she said. "Being prejudiced is being biased or one-sided—forming an opinion without all the information. It is a form of 'pre-justice' indeed.

"You can bet your last dollar on the fact that when, and if, I get out of here I'm going to bust my butt to get back into college. And I'll damn sure finish if it takes me ten years!" It almost did.

The bus pulled up to the corner at the end of our company street and disgorged five or six soldiers. Barry and I took our little bags and headed for the Orderly Room to report for duty.

"Here we go again," I said as we went in the door.

CHAPTER TWENTY-TWO

Squad tactics require that each member knows what the other fellows are doing, and—just as importantly—where they are located. This type of cohesiveness is, of course, supposed to extend right on up to the platoon and company levels. Regimental and brigade tactics were well beyond our expected knowledge, and were under the authority of field-grade officers like Majors, Colonels, and Generals. These men were way up the chain of command and ones we seldom even set eyes on.

It was just getting light one morning as we stood in formation in front of our barracks. The weather forecast for the day was temperatures in the uppper fifties so we were not wearing our field jackets.

"Today, young ladies, we are going to engage in a company-sized maneuver where we will defend a simple fortified position. 'Simple' in this case means you-all will build it, or, rather, dig it, and occupy it for only a short time." *I wondered to myself how many times Sergeant Talbot has been through this exercise and for how many years.*

Trucks delivered Company B to the training area, where we unloaded and were greeted by an instructor. The immediate terrain was desolate—a series of small hills or ridges almost devoid of vegetation. Every so often a small pine or oak poked its head up to a height of about three feet, rising above the occasional log that lay on the ground. Several hundred yards distant the land was more timbered.

"Each platoon will establish a defensive position along that hill and will be in close proximity to the other platoons. You will dig an entrenchment that is approximately two feet deep, piling the dirt in front of you facing that tree line over there." The Sergeant pointed in the direction we faced. "Each squad's entrenchment will join that of the next squad's. The ground is soft where you'll be digging. This is certainly not the *first* time a company has defended that hill." The Sergeant was holding a long stick and was scribbling in the dirt with it as he spoke.

"This don't sound bad to me," McCoy said. "Looks like all we got to do is go up on top one of them little hills, dig out some sand, and sit there all day looking down there." He pointed to the trees.

"How many times have we thought something was going to be easy and it turned out to be pure hell?" Lee said, as he slung his M-14 over his shoulder.

Once we arrived at what was to be our defended position, we were ordered to dig a shallow trench that, when the dirt was placed in front of it, would provide some protection from enemy fire. Mahoney and I were on the left flank of our squad, and we could see a relatively open space for about three hundred yards in front of us.

"You M-60 guys need to position your weapon so that you can protect this side of the line." The Sergeant handed me a piece of stiff cardboard and a pencil. "I want you all to walk to that tree line and map obstacles between here and there that could be used as concealment by an enemy. Show any logs, gullies, or clumps of weeds or bushes within a forty-five degree angle area. Your gun has to be able to 'sweep' over that much ground. The other machine gunners will do the same thing."

The M-60 is an awesome weapon. It weighs about nineteen pounds and can be fired from a stationary position using a heavy tripod, or on the move with a much lighter folding bipod to provide stability. The gunner can actually fire the weapon while he's standing up. The M-60 has an effective field of fire of about 3,600 feet, with a maximum distance of about 2.3 miles. Approximately 550 bullets can be fired per minute, but the barrel must be changed frequently at that rate to avoid jamming. That is something the shooter must be aware of in combat. And, weapon jamming is a situation that infantry soldiers have always faced in battle.

"Come on, Manooch. Bring your paper and pencil and

we'll get this little detail out of the way." We walked all over the terrain in front of us and did as instructed—marking all potential "hides." That afternoon was one of the most enjoyable that I spent in training at Jackson. Most of the day and for part of the early evening we stayed in the trench either dozing or looking in front for something to happen.

We were provided with a password that we used to challenge anyone observed approaching the trench. This happened several times that evening to test our readiness. Passwords are in two parts like "pork" and "chop." Some passwords seem about that silly. The first word is called out and the proper response is the second word. We used passwords when we were on guard duty, which included guarding the ammo bunkers on Base. On those rare occasions we were provided with live ammunition. We concluded the day's exercise and returned to barracks around 1900 hours.

The next morning we started another training exercise.

"How long you guess we'll be out here?" Lee asked as he adjusted his field jacket. It had turned cold again like it frequently does in early spring in the Carolinas. We were once again outside and ready to receive training instructions.

"Probably all day. They have to show us different types of "booby traps." They're the things soldiers stumble into when they're in combat. From what I've seen in old World War II picture shows, a fellow could be just strolling around in the woods, everything looking as peaceful as can be, and 'BOOM' he's tripped one of those things and he's dead or lost his legs."

"Don't be so damn morbid, Plank," Johnson said as we filed into the bleachers for yet another time.

That day we were introduced to the lengthy history of using booby traps, which went back several thousand years. We were shown examples of both primitive ones and complicated ones used today.

"This is the M-18A1 Claymore directional antipersonnel mine. It is named after a large Scottish sword." The Sergeant held up a rectangular shaped mine that was curved. It looked to be about sixteen inches in length and about eight inches high.

"The Claymore fires steel ball shrapnel out to a distance of about one hundred yards in a sixty-degree arc in front of it. These things can be used to ambush the enemy. They can be fired in the 'uncontrolled mode,' using trip wires and sensors, which are set off by enemy infantry. However, the most effective means of firing these things is by the 'controlled mode' where an observer waits until the enemy is concentrated in the killing area, and then he sets it off."

Plank was sitting there with a disgusted look on his face. "I hate chicken shit like that."

"What the hell are you talking about Plank?" McCoy responded.

"To me that's fighting dirty. Not even man enough to face your enemy, but rather let some damn booby trap kill 'em. When I fight I want to see them face to face."

"Plank, what in the hell do you think snipers, artillery, airplanes, mortars and a whole bunch of other shit do?" Mahoney had jumped into the argument. "When most soldiers die in war they never see it coming. You think we still live in the days of duels or knights charging each other on horseback?"

"Naw, but I still believe it's chicken to hide way off somewhere, push a button, and let that do the work for you."

"Our job as soldiers is to defeat the enemy any way that we can. If I have to kill somebody I'd rather do it on my conditions and not his. The object of war is to win—not play fair."

It was a losing battle and Plank knew it. However, he was still muttering about "chicken shits" when we boarded the trucks and drove away from the training area.

CHAPTER TWENTY-THREE

Considering the fact that the fall and winter had been so darn cold and wet, we were naturally glad to see an early spring, which held promise of being drier than usual. The sand hills with their scrub oaks and pines were drying out. There were some areas on Base that had much larger trees—we had passed them on our way to different training areas—but where we usually spent our time had a tendency to look more "trampled down." In fact, Fort Jackson covered thousands and thousands of acres, but we saw only a fraction of it while we were there.

"I heard that the Cubans are starting forest fires," Toby Lee casually remarked one morning as we stood outside the barracks. The soft-spoken, rather off—the shoulder—comment, got our attention. Moody asked the question each of us wanted to ask.

"What the hell do you mean when you say 'the Cubans are starting forest fires?'"

"I mean exactly that. *The Cubans are starting forest fires on the Base!* When they get tired of training, which I hear is all too often, they set fires in the woods so that they can stop training."

"How does that get them out of training?" I asked Toby.

"They sit back and wait until someone else puts the fires out—that's how. And, you can 'bet your bottom dollar' that that *someone* could easily be us next time."

Toby Lee was right, and it wasn't long after that morning before our infantry company was learning more than we ever wanted to know about fighting wildfires.

Although I was certainly not having a raucous time at Fort Jackson I was beginning to be proud of what we had accomplished in just the past four months. We had traveled a long way, in a relatively short time, on the road to becoming men. You could see it in the way we marched, and in the way we followed orders from the NCOs. We griped far less, and responded quickly to orders that were given to us. We realized that we were an important part of a much larger force of men and women who represented the United States as soldiers in the Army. This positive attitude followed us as we continued to train.

The typical scenario for our company-sized maneuvers was a simulated enemy attack during the daytime. When this occurred we took immediate cover, regrouped and adjusted, and then went on the attack. The concept behind this was to not stay in a defensive position for long, but rather to repel and pursue the attackers. These maneuvers seemed very real to us, even though we used blank ammunition while our instructors hurled smoke grenades and artillery simulators into our ranks. In all the confusion of shouting, shooting, and booming noise, the biggest fear we always had was to be cut off or isolated from the main "friendly" force. We could only imagine how this fear must be overwhelming during actual combat. It was stressed over and over that an isolated combatant can more easily be killed or captured.

"Mahoney, hurry the hell up, you big son of a bitch," I yelled as we ran through the scrub oaks. Barry was a few yards behind me and ran bent over, kicking up sand as he carried two ammunition cans for the M-60.

"Damn it, Manooch. Slow the hell down! We're supposed to be on the left flank, not out in *front* of everybody." I slowed, looked to my right, and caught glimpses of a couple of our guys moving at the same rate and direction as we were.

"Come on Mahoney, get up here with me. We sure as hell don't want to get separated. This thing can't fire without ammunition in it."

One of the Sergeants behind us blew a whistle and we dropped to the ground in preparation for firing. Mahoney crawled up to me with his butt stuck up in the air. I guess it's not easy for a six-foot-five guy to crawl very close to the ground. I adjusted the bipod and threw the breech open and Barry inserted the ammo belt from the left side. Hundreds of 7.62-mm blanks were coiled in the can and fed into the breech by a metallic belt, which broke up into many pieces when the M-60 was fired. Each round had shiny new brass and a small red tab in the end—indicating that it was a blank. A second blast on the whistle and all hell broke loose as the woods erupted in small arms fire from our entire Company.

I fired the machinegun from the prone position, and the air was filled with small metal clips and "spent" brass flying out of the breech. Suddenly the ear- splitting whining, whistle-like sound of an artillery round simulator caused us to hunker down in the sand. We hugged the ground even closer when the simulator emitted a loud "BOOM" almost on top of us.

"Damn. Those things always scare the hell out of me," Mahoney groaned as he peeked from under his steel pot.

"Just be glad they're not real," I replied as the firing began to subside.

"I wonder how we did? We're obviously not wounded, dead or captured."

"The referees must consider things like how well we defended ourselves, and how we were aligned during the attack," I said as I brushed sand off my knees and elbows. From several hundred yards to our rear we heard the order to regroup, and we immediately started to jog back towards the trucks.

Talbot's monkey was sitting on the ground like a raccoon and was leaning back with its front legs between his back legs. He looked directly at us with his black shiny eyes. There was no question in our minds that the monkey loved where he was, and what he was doing.

"Come on. Form up on me," Sergeant Talbot shouted

as the monkey jumped up and down, and then ran up his leg and sat on his shoulder. "Got your entrenching tools with you I hope?" he said sarcastically. "You better have—because that's what you're going to fight a damn fire with."

"I knew it. Damn. I just knew it," Toby Lee said disgustingly as we climbed into the waiting trucks and took our seats along the sides. "Those sons of bitches have pulled us into it now," Toby continued to rant as we drove away.

We smelled the smoke long before we reached the fire. It smelled like wood, of course, and as we got closer we saw that the smoke was white, with ashes floating down every so often. As we unloaded we felt the heat for the first time, and could hear the crackling from flames in the nearby underbrush.

"This is one time fellows when you *must not get separated*. This ain't war games here. Be very careful as you work to keep contact with the guys on either side."

"Damn, I believe Talbot is being compassionate," Mahoney said as he unfolded his entrenching tool and screwed the handle tightly in place.

"I wouldn't go quite that far," I replied as I looked at the smoke, which seemed to be getting thicker by the minute.

"For today's little exercise—and for only today—you may work in your t-shirts and you do not have to wear head gear. Since we are the first on the scene, we have to use what's available to us until we receive some professional help with this thing." Sergeant Talbot lined us up and we walked down a sand lane where the fire was creeping up to us from the right.

We were lucky at this point, because the flames were close to the ground—consuming dry leaves, pine straw, and anything else in the way. Behind that initial fire line the flames were larger and were engulfing the small trees.

"Your job is to break the progress of the fire," Talbot shouted. "And we can attempt to do that by shoveling sand on it and by removing any combustible debris from its path."

We went at it hot and heavy, using our entrenching tools to beat at the flames and to shovel sand onto them. In no time at all, my shirt was wringing wet from sweat caused by the heat of the fire and from physical exertion. In fact, we were caught up in the apparent urgency of the situation, and we were going at it too hard. I looked over and Mills was sitting on the ground struggling to get enough air into his lungs.

"Dem damn Spics. Chicken shits. Dey started the fire, and then dey 'high tail' it off somewhere and take it easy while we're in here busting our butts. Shit. Something ain't right about this."

"What's worse," Mahoney replied, "is that no one's going to say anything about it because the old U.S. of A. doesn't want to alienate that 'Free Cuban' voting block down there in South Florida. *That* would not be politically acceptable."

"Suppose one of us gets killed out here doing this?" McCoy said as he wiped at the sweat running down his face.

"They'd just say you died in the line of duty by a natural act of God." Mahoney continued to amaze us with his knowledge about almost any situation. Or, we were too gullible one way or another. The heat and smoke were almost overwhelming now, and as we returned to work we saw that the fire had crept to the edge of the sandy lane.

"Don't let it 'top out.'" Sergeant Talbot shouted. "If it jumps this road there will be no way that we can hold it back. There are some big trees over in the next bottom to fuel the fire."

That brought us back to action and we returned to slapping at the flames on the ground and clearing potential fuel out of the way. It was obvious, however, that this effort was too little, too late, and was doomed to fail. We were all exhausted.

"Reinforcements are on the way, fellows." Johnson was pointing down the road where the sounds of sirens were wailing through the scrubs. Several minutes later two fire trucks pulled up and firefighters disembarked. Some had backpacks that contained water. Others grabbed shovels and similar tools that could be used to slow the flames while the water was being applied. These fellows clearly knew what they were doing. Thirty minutes later it was all over. The smell of smoldering wet ashes and the occasional whiff of white smoke was all that remained of the Cubans' fire.

"Damn. Would you look at that," Lee said and we turned and followed his gaze. A beautiful red cardinal was sitting on a limb of a partially burned pine—and he was singing his heart out.

CHAPTER TWENTY-FOUR

Our stint in AIT was drawing to a close, and most of us were going to continue infantry training in a Basic Unit outfit, which was also located at Fort Jackson. Johnson and I took advantage of three-day passes and we went home. These trips were technically not allowed because of the limit on distance that could be traveled. Raleigh and Archer Lodge were too far away for a three-day pass, so these trips were exciting from start to finish. We left on Friday afternoon and had to report back by Sunday afternoon. Therefore, much of the time was spent on the road; the rest was spent visiting our girl friends. There wasn't much time to get rest, so when we reported back we were usually worn out.

Sometimes we just spent the weekend days of Saturday and Sunday on Base. This was free time when we could sit around, shop for a few essential items, and eat more leisurely in the mess hall. Four or five of us formed a little social group. We stayed together most of the time, and got really close. One afternoon our conversation touched on the topic we hardly ever thought about or mentioned.

"Did you ever stop to think how it would be for us in combat," Johnson asked.

"What do you mean? It'd be tough on anybody in a war," Mahoney said from his seat on the steps to our barracks.

"No. You don't get what I mean. I mean how would it be for the five of us, as close as we've become, to go into combat and one or more of us get killed or seriously wounded. It would be very hard on the rest of us. Remember back in the Second World War when Reserve or National Guard units from the same little towns were sent off together and how tragic it was for the families and the communities when some were killed?" Johnson was right, and each of us was thinking about examples when this had occurred.

"You know there was a little town in Virginia that sent its sons off to war and they landed in Normandy on the 6th of June, 1944, as part of the Allied Invasion of Europe. Many of those boys were killed on the first day." Mahoney had his thinking cap on and he was on a roll. "And do you guys remember the 'fighting Sullivans' who were in the Navy fighting the Japs in the Pacific. They were all on the same ship when it went down in battle. When the Mother of those boys got 'the call' she knew what it had to pertain to so she asked, "which son was killed?"

"They *all* were, ma'am," was the response. Mahoney had our undivided attention and we sat there real quiet for a few moments.

"A little closer to home," I said, "was when Sergeant Talbot was telling us some of his war stories. Remember that evening when he came in the barracks, sat down on a bunk and just talked to us like we were his buddies?"

Johnson came over and sat next to me. "Remember? Hell who wouldn't—considering how he's always giving us hell, and besides, that was the first and only time we'd seen him without his damn monkey."

"I don't mean that. I mean do you remember how serious he was when he was talking about the Korean War? His eyes were moist when he told about training young men—he said 'boys'—for months, and then leading them into combat, and how terrible it was for him when he saw one of them wounded and dying. He said it so softly, but I'd swear he said it was as if he'd lost a son of his own."

Mahoney got up and started to walk away. "Yeah. That's one hell of a responsibility, and that's why he's so hard on *us*. Down deep he's not mean. He just doesn't want to lose any more of *his boys*."

One final thing we had to do in AIT was "show our stuff" to recruits by starring in the "Infantry—Queen of Battle" firepower demonstration. The entire Company, in full battle gear, was loaded into covered two-and-a-half-ton trucks for the trip to the firing range. McCoy, Mahoney and I were sitting next to each other. "Remember that day months ago when the three of us were detailed to the firing range?" McCoy was saying aloud what we were thinking.

"Who could forget that?" Mahoney answered. "That was the day the two of you were almost burned alive stirring napalm."

The trucks slowly worked their way across Fort Jackson and then turned off the pavement onto sandy roads to ferry us way out to the boondocks. "Hell, I don't remember it being this far out here," I said as I peered out the back of the truck.

"Well you ought to. You've been to this range two times. The first time was when we went on that work detail, and the second when our Basic Training Company was sitting in the bleachers watching the show. I guess you were in some distress those times not knowing what to expect. Today it'll be a more relaxing visit." Mahoney was probably right.

"I want you to explain something to me, Mahoney. You're so damn smart and everything. How can a sand road make dust like this?" Johnson pointed at the truck behind us, and you could hardly make it out because of all the dust swirling behind our truck.

Barry thought about this for a few seconds and then offered an opinion. " I don't know about the dust, but I do know that military training bases are located where nobody else wants to live. The land is no good for farming or logging, and you sure as hell wouldn't want to plop a housing development out in one. Have you fellows noticed that out this way—and off the Base—there are very few houses, and those you see look like something on a tenant farm or straight out of *God's Little Acre*?"

I remembered, all right. And I thought about the time while being driven out to a work area I saw a woman hanging clothes on a line, the wind blowing, billowing out her washing and probably blowing dirt into it, and the yard devoid of any trace of vegetation. Yards like that made me think that chickens had been let loose in them. Chickens'll eat anything—pecking the ground completely clean. *How do people living like that have enough drive to get up every day and go on? What chances do the children have if they're raised up in that kind of poverty? At least they don't live in some big city ghetto where folks go to sleep at night listening to the sound of shooting on the streets.*

The truck hit a bump, which stirred me from my daydreaming, and it slowly turned off the road with the other trucks and pulled into a parking area behind the bleachers.

"Would you look at dem dumb asses?" Mills said as he pointed to the soldiers seated in the viewing stands. "Those poor sons of bitches are just getting started in Basic. Look how new their fatigues look."

We all turned and stared at the recruits, who indeed had on uniforms that were a dark olive drab, and very much unlike our faded ones. I said: "How in the hell would you like to be sitting there just *getting started* with Basic? Knowing what we know and what we've been through. I'd jump up, climb down from there, and run like hell for the fence."

"Manooch, you're too damn dramatic," Mahoney said as we were called into formation. We lined up smartly and waited for the commands of Sergeant Talbot.

"I want to give them a demonstration to remember," Talbot yelled out. "Show them what only four months of hard training will do. Each one of those recruits will leave here today knowing what they're capable of becoming.

Let's go, Infantry!"

Of all the NCOs we had at Fort Jackson, Sergeant Talbot was the one whose name I remember best. Most of the other names have been forgotten but I still can see their faces. Remembering Sergeant Talbot's name could be because of his small size, or his monkey, or a dozen other things, but it was probably because he really cared for us. Not in a soft way. On the contrary—he was extremely hard on us. We saw that the first day we were assigned to his platoon. We were his troops—his boys—while we were in AIT and he wanted to make sure we were trained as well as possible in the event we had to go to war. All of us knew this now, and respected him for it.

We were in good spirits as we thought about the continuation of our training cycle and getting closer to the day we could go home. However, some of those in my platoon expressed regrets that Sergeant Talbot was not going with us over to the Basic Unit Training Company. We'd even be happy to see his monkey come along.

CHAPTER TWENTY-FIVE

The company area that we were assigned to for the final leg of our stay at Fort Jackson was different from previous ones. Our new barracks in Company D, Fourteenth Battalion, Third Regiment was located in an area where there were actually tall pine trees growing very close by. The topography was also different. It was more hilly. There was some rise in elevation, which, in combination with the trees, gave a more welcoming appearance. The main entrance to our barracks was right beside a sidewalk lined with pine trees. The company street—the troop formation area—was behind the barracks, which for us was unusual. As with my previous assignment, my bunk was on the lower floor, and those of us that had been together in Basic and AIT were again placed in the same company.

There were some new faces, of course. One of those was Barry Gutmann, a Jewish boy from Dayton, Ohio. Gutmann was about my height, although lighter in weight, and his cheeks appeared to be always chapped and red. He liked to talk, and it didn't take him long to tell us all about himself. " You fellows ever hear of the Mafia? Probably not, if you're from down South. But up North in the larger cities the Mafia is often not only present, but is involved in politics and economics—it runs things. I know because my father owns a tire business in Dayton and the Mafia pretty much tells him what to do."

"Well I'm from Ohio too, and I haven't heard much about criminals being in charge of things where I come from," Mahoney responded. "But I guess Dayton is different."

"It is different," Gutmann replied as he took a map of Ohio from his wall locker. "If you'll look at this you can see all these industrial areas where most of the working men and women are employed by large steel companies. The unions are strong, and in some instances, have been taken over by the Mafia."

"Well, what in the hell is the Mafia?" Johnson asked as he looked at the map spread out on my bunk.

"The Mafia, my country friend, is a powerful, ruthless, organized crime group that originated in Sicily and worked its way to the United States in the nineteenth century." Barry really had our attention now and Toby Lee joined the group.

"Haven't you seen the movies where those guys in long overcoats and fedora hats ride around in old cars and shoot at people with Tommy guns?"

"I thought all that stuff died out in the thirties when Al Capone was sent off to Federal prison," I said to Gutmann.

"It did for a while, and it's certainly not as strong and influential as it once was. However, there are places—New Jersey and New York—where it has remained strong or has established a foothold like it has in Dayton." Gutmann was getting worked up now and his cheeks were almost crimson.

"Better calm down Gutmann, or you're going to have a heart attack."

"Don't sound so concerned Johnson. I'm just fine." Barry put the map back in his wall locker, and the rest of us resumed what we were doing before being interrupted by the Mafia lecture.

Basic Unit Training was exactly what it claimed to be, and involved intense training at the platoon and squad levels. It also required us to spend much of the time in the field, and we were soon engaged in maneuvers that lasted from twenty-four to seventy-two hours. We were lightly equipped, and spent the nights in foxholes as opposed to setting up tents. All of this was intended to make us more mobile so we could respond to a situation—either offensively or defensively—and then quickly move on.

It was now early April of 1963. The weather was actually pleasant, and most of us had become fully accustomed to military life. We were still required to serve on work details as assigned, which were in addition to the normal training schedule. One afternoon the Sergeant called me out and ordered me to walk guard duty at one of the Leesville Road rifle ranges. Leesville Road is way the hell out in the boondocks, and I remember thinking *who in the devil would want to steal anything or destroy something all the way out there off Leesville Road?* It really didn't make any difference what I thought—I was going to walk guard duty out there all night.

A three-quarter-ton truck carried me and two other fellows out to the ranges. We were sitting in the back of the truck and each of us was armed with a loaded M-14 rifle. We wore steel pots, and since rain was in the forecast we carried our ponchos—rolled up and on the back of our belts. "Good Lord, it's dark out here," one of the guys said. He was right about that, as there wasn't a light to be seen.

"Okay boys, one of you get off here," the Sergeant said as he leaned out of the window. "There's a field phone on that telephone pole over there, and I want you to check in every hour."

"What do we talk about Sergeant?" the fellow asked as he climbed out of the truck.

"You don't *talk* about anything, you dumb ass. You merely call the Officer of the Day and give him your name and where you are. He'll know that you're okay, your position is safe, and you're not overrun by bad guys."

It had started to rain as the truck drove off and we could see that the newly-posted guard was unrolling his poncho. The two of us were sitting across from each other as we proceeded to the next drop-off point.

"You know this is gonna be one hellava long night, and it'll probably rain the whole time. What did they say the password was? Something about 'rocky.'"

"It's 'Rocky Mount,'" I replied as I tried to appear unconcerned and somewhat bored with the whole thing.

"What if somebody comes up to us out there and they don't give the right password? Are we supposed to shoot them?" The poor fellow was dead serious and was obviously working himself up over nothing.

"Look. Use your head. Do you think I would shoot a person out in the middle of nowhere in the dead of night just because he walked up to me while I was guarding some damn rifle range that's not worth shit to nobody?"

"What if he tried to take your gun?"

"Then that's different. You'd have to defend yourself and you couldn't let him have your rifle." *Jesus Christ*, I thought, *what will he come up with next?* I didn't have to ponder that long because the truck pulled off the road and I took the opportunity to jump down. I figured the sooner I got away from all the jabbering, and got this detail over with, the better.

Once on the ground the first thing I did was walk over to the pole where the phone was supposed to be, and seeing it in its proper place gave me some reassurance. The rain was starting to come down pretty heavy, so I unrolled my poncho, slung my rifle over my shoulder, and put my poncho on. I then began to walk my assigned area. It took me about five minutes to walk from one end of the firing stations part of the range to the other. That's in the front of the range where the parking area is located. There were three main wooden structures—a tower where the range officer observed the firing; a storage building where range supplies were kept; and an oversized latrine. The wind had picked up and I could hear it moving the tops of large pine trees, which were swaying back and forth. Although it was lonely as heck, quiet time by myself was a relief because there was nobody in my face yelling at me all the time.

Time passed by quickly and it wasn't long before I had to report to the Officer of the Day. (I don't remember wearing a watch while I was at Fort Jackson—someone else kept time for us.) Much to my relief the call went smoothly and I resumed my patrol, which included an inspection of the three buildings. When I got to the latrine, I decided to go inside and give it a try. For a bathroom, the place was huge. It had enough holes to accommodate about twenty soldiers at a time. The "throne area" was made of wood, slightly elevated off the dirt floor, and consisted of two lines of holes, ten on a side. During times of heavy use, troops sit back-to-back positioned over each hole. My "urge" was getting stronger so I prepared to mount the throne. I now know what girls and women in dresses and underpants have to go through just to sit on the "John." I had to drop my pants and drawers, and then lift my poncho up so that I wouldn't be sitting on it. My hood was up over my helmet and water from the rain was trickling down my face and getting in my eyes and mouth. I thought I felt something tickling my upper lip, and it didn't seem to be water. When the realization hit me that I was being visited by some type of bug, it was too late. Whatever it was had stung or bit the living hell out of me, and I instinctively reached up and swatted it to the floor. It was a large spider—and I stomped it with a boot.

Almost immediately my lip began to swell and throb and it hurt like hell. At first I tried to put it in the back of my mind like *this will be okay in a few minutes, don't get panicky, just be calm.* However, it wasn't okay after a few minutes, and when I felt my lip I couldn't believe how big it was. A large bulbous portion extended down over part of my mouth. I decided it was time to get on the phone. My speech was obviously altered by the swelling.

"Hello thir. This is Private Manooth, thuard at tha Leesthville Rangth Number thrwee. A thsida bidth my lipth. Thud I be relievth?"

"Manooch, what in the hell is going to bite you next? A damn alligator? I personally don't give a damn if you die. However, stay where you are and an ambulance will come and take you to the hospital. A new guard will be brought to relieve you." I hung up and waited.

Not long after the call was made an ambulance picked me up. Once in the emergency room, I was seen by a doctor who was a Captain. That's what most medical doctors were. "What type of spider was it, Manooch?" he asked as he and a nurse took my blood pressure and used a stethoscope to check my heart. They seemed satisfied with the results, but neither of them could keep their eyes off my pulsating lip.

"Whath typth oth thspida wasz itdth? Howth the hellth doth youth thinkth I knowth? Ith sthopmth the shith outh od ith."

"All right Manooch, calm down. Nurse, put him in Ward Two and we'll observe him overnight. If everything goes like I think it will, we'll release him in the morning."

"Thankths, Doc," I said over my shoulder as I was rolled down the hall looking pitifully unfit for duty because of the badly swollen lip. I was discharged the next day, and returned to my unit.

CHAPTER TWENTY-SIX

One of the most unforgettable characters I encountered at Fort Jackson was Dillon "Mud Flap" Shepard, a boy in our unit from Tennessee. He was from a National Guard unit back home, and like those of us who were on active duty for only six months, was referred to as a "six month wonder" or "weekend warrior." Mud Flap stood about five foot-four, was always smiling, and looked a hell of a lot younger than he was. His knuckles had words tattooed on them. "L-O-V-E" was on his left fist, and "H-A-T-E" ran across his right. He said life was a continuous battle between good and evil, and to demonstrate this he'd clasp his hands together, interlocking his fingers, and struggled until one fist was victorious over the other. Most times it was "L-O-V-E."

Mud Flap was raised back in the mountains where folks went to town only once in a while. We guessed his schooling had been, at best, irregular, and we doubted that he'd ever set foot in a doctor's or a dentist's office. This assumption was at least partially proven one day when we had to have an end-of-cycle physical. I remember the day well because the dentist was from Big Spring, Texas and that's where I was born. He's the only person from Big Spring that I've met before or since. Mud Flap was right in front of me in line.

The dentist was using a metal pick, one of those sharp-pointed curved ones, to do a quick inspection of each trainee's mouth. The examination of Shepard proved noteworthy, because the dentist was able to dislodge three or four rotten teeth with only moderate effort. I can still hear the sound made by each tooth as it was plopped into the metal trashcan. Shepard was the only reservist or National Guard member in our company who was fitted with a new-and free-set of dentures before he left Fort Jackson.

Barry Gutmann came into the barracks one day grinning from ear to ear. His faced was flushed—as it always was when he got excited.

"You'll never guess where I'm going in two weeks," he said, barely able to hold his news.

"Why don't you just simply tell us and save a lot of time," Johnson replied, showing little interest. Several of us stopped what we were doing so we could pay more attention to what was being said.

"First let me tell you where I'm going, and then I'll tell you whose guest I'll be. I'm going down to Athens, Georgia for The Masters." He relished the significance of his announcement, and waited for it to sink in. It didn't.

"What the hell is The Masters?" Lee asked. I don't believe any of us knew what it was.

"*The Masters* is a professional golf tournament. One of the most well known in the world—that's what." We remembered then that when we first met Gutmann there was some talk about him playing golf.

"I'm actually a professional golfer, although I have to admit that I don't yet play with the big boys—the truly great golfers in America. When I entered Basic here at Jackson the golfing aspect was reported in my background materials. You do realize that right many professional athletes—like football players—go through Army training here. It's not that uncommon."

It was amusing to see us looking at each other, knowing that we were in the presence of greatness, and that greatness was in the form of a chubby, red-cheeked Jewish boy.

"Now what's really unusual is that I have been asked by General D'Orsa to be his guest at the tournament." That was way, way beyond our comprehension. Hell, a Captain, usually a clergyman or doctor, was the highest-ranking officer we ever saw up close. While we were left to our training, Gutmann was tooting off to watch a golf tournament with the Commanding General of Fort Jackson. We were sore at the thought, but we weren't mad at Gutmann. Who wouldn't jump at the chance to spend a few days off Base.

With Gutmann off to Georgia with the General, we still had good old Shepard to keep us entertained in our little corner of the barracks. He could come up with something all the time. Mud Flap had a saying he used when he was asked where somebody was, and he said it frequently. I thought it was the most common, asinine thing I'd ever heard. And, most of us would've never said anything like that around our folks back home. However, I got to thinking about it and what the saying was derived from. Let's say you lived way back in the Tennessee hills—or the North Carolina, Georgia, Virginia, West Virginia, or Kentucky hills for that matter—and your house was a small three-room cabin with a garden, and chickens and pigs roamed free out in the yard. There was no running water or bathroom inside the cabin, and you had to use an outhouse that was fifty feet or so from your back door. Let's say that on one cold, late spring night you had to "go" real bad, so you got up and headed for the outhouse. You made your way barefooted through the collard patch and walked under the chickens that had all gone to roost up in the trees. Just as you reached the outhouse you slipped in the mud and fell flat on the ground. The pigs were right there. You can see now, how Mud Flap could have been right when he said about the absentee: "He went to shit and the hogs ate him."

Right after Gutmann returned from his golf tournament the Company took part in a seventy-two hour maneuver. We were trucked way back in the woods and were placed in a defensive position to protect our headquarters and communications center from infantry attacks. Most of the time we stayed in foxholes. Gutmann and I were in the same one.

"Look what I brought along," Barry said as he proudly showed me a candle and some wooden matches.

"We're supposed to be in a black-out situation, you idiot. You'll have us called out for sure." I was leaning back against the sand wall of our hole. That's all we could do in there passing the time away. Just sit or kneel and peek out over the top to see if anybody was coming.

"Nobody's going to see the light because we can cover this hole with our ponchos. We'll wait until we've eaten supper, and then we can come back here, arrange the ponchos, light the candle, and you can crawl outside and tell me if you can see any light."

"Why are we doing this in the first place?" I asked. "Can't we just sit in the dark like everybody else? It reminds me of something little boys would do just to see if they could get away with it." I thought back when I was about twelve and Butch Royster and I had taken Billy Bernhart's boat out on Boone's Pond. We spent the night out there too scared to go to sleep. We had a red Boy Scout kerosene lantern with us, and it provided some light—similar now to the dim light in our foxhole. I woke up coughing and with eyes burning. It took a few moments for me to realize where I was. The foxhole was full of smoke, which seemed to be coming from Gutmann's web belt—his fabric canteen cover to be more exact. "Damn. Wake up Gutmann! You've set the place on fire." We both scrambled around to cover the smoking mess with sand, and then flapped the poncho up and down to vent the smoke.

"What the shit's going on over here?" A Sergeant from

the next company was taking a shortcut through our bivouac area. We peered out from under the poncho and looked up at him.

" One of these simulators went off and caused all the smoke," I said, hoping my startled face would cover up the lie.

"Well, get it out! You fellows looked like a couple of Indians sending up smoke signals."

CHAPTER TWENTY-SEVEN

One of the biggest mistakes we made while in Basic Unit Training was showing Mud Flap how to light farts. This intellectual endeavor involves a combination of physics, chemistry, and that trait only displayed by human beings—out-and-out crudeness. The setting for this great revelation was one evening after we returned to barracks from the mess hall. A game of poker was in progress when Gutmann stood up, bent over, and quick as a flash, literally, whipped out his lighter and fired off a respectable fart. Mud Flap, who was lying in his bunk nearby, sat up, mouth agape, and stared in *total disbelief.*

"*What the hell was that*?" he said with a silly sheepish grin on his face.

"That, my friend, was an example of torching off internally, self generated methane gas," Gutmann replied as he put his lighter in his pocket and nonchalantly sat down to resume his place in the card game.

"You mean farts can burn? Here, let me try it," Mud Flap said as he came over to join the group. He borrowed Gutmann's lighter and then worked for several minutes to summon an appropriate expellant, and then lit the lighter and pooted. Nothing happened.

"No, no, Shepard. You have to hold the lighter *right against* the seat of your pants or the flame won't ignite the gas. Here, let me demonstrate again." Gutmann got up and repeated his prior performance, which exhibited his obviously significant experience.

We could tell that a strange transformation had occurred in Mud Flap's mind. It was almost scary the way he smiled, and with a wild gleam in his eyes he grabbed the lighter and went back to his bunk. For days afterwards, Mud Flap would sneak back to his bunk during free time. He lay on the bottom bunk on his back, balled up, and with his toes grasping the metal springs of the upper bunk. He spent hours perfecting his technique so that pretty soon he had it down to a science.

The utterly delightful concept of flatulence ignition became an obsession with Mud Flap and before long it was all he could talk about. "I'm going for a record," he proudly proclaimed one afternoon.

"In order to do that, you know that everything has to be just right. Can't be any moving air; there has to be plenty of high quality gas; and the lighter has to be placed exactly right and at the split second that it's needed." Gutmann was providing what he considered to be wise insight and encouragement at the same time.

That Friday afternoon, as we were lined up outside the Mess hall, Shepard was shamelessly soliciting up and down the line. " It's Friday night and you know what that means. We're having fried fish, coleslaw, and *baked pork and beans*. I'd be proud to take the beans off your hands if you don't want them." Several guys accepted his offer and before long beans were piled high on his plate.

Mud Flap staggered back to the barracks after supper, and his belly stuck out like he'd swallowed a cantaloupe. After a brief rest he decided it was time to go for the record—something he had come to cherish more than anything in his life. We gathered around his bunk and eagerly awaited his try. He assumed "the position," but his first attempts produced only what you might say were good quality flames—each about three or four inches in length—about average. The disappointment was written all over Mud Flap's face as he dropped his drawers and lay back on his bed to try it again.

"You're not going to do that *buck naked,* are you?" Johnson asked in utter amazement—a feeling shared by all of us.

"Hell yes. I'm going for it all this time." And after much grunting and straining Mud Flap cut loose with a fart that would've registered on the Richter scale. There was a loud *Whooosh* and a glow of blue light filled the barracks as the gas ignited in a two-foot stream. Unfortunately, it did not immediately extinguish as usual, but continued to burn for two or three seconds.

Shepard cried out like a hurt animal, which caused our momentary laughter to cease—replaced by feelings of pity and compassion. "Is the Mess Hall still open?" one of the fellows inquired. "We need to get Mud Flap some ice as quickly as possible." No one on our bay slept that night, as we lay there in the dark listening to the poor boy whimpering as he held an ice pack tightly against his rear end.

Next morning in formation, and after Mud Flap was allowed to go on sick call, the Sergeant told us: "I don't believe I have ever had a trainee go off to the hospital with a burnt ass hole."

The end was drawing close. When our time of active duty had dwindled down to fifty-two days, we got a deck of cards and tore one up when a day was over. We had a ceremony where five or six of us gathered around for the actual tearing up. I don't remember in what order the cards were arranged. However, the numbered cards started off with the twos in one suit and worked up. Then another suit was selected and the process repeated. The last card was the queen of spades. Maybe because one of us remembered that sign: "Infantry—Queen of Battle."

During our last week we had to have a physical examination before being released back into civilian life. "They've injected, withdrawn, and probed my stuff to the point there can't be nothing left," Toby Lee said as we stood in line, stripped down to our swing easies.

"Here. Each of you guys take one of these cups and pee

in it for the doctors."

"That orderly is so 'full of piss and vinegar' let him do it," I whispered as we each took one of the little paper cups.

"I can't pee," Lee said with a strained sound to his voice.

"Me neither," Johnson replied as he peered at his empty cup.

"Well I sure as hell can," Gutmann proudly stated. "You guys give me those things and I'll do it for you." He did, and we left the infirmary much relieved.

Several days later, on the queen of spades day to be exact, we stood in what was to be our last official Company formation. When given the dismissal order, we whooped and hollered and jumped up and down in pure exhilaration.

"Manooch, Lee, Johnson, and Gutmann, hold on there a minute," the Sergeant shouted over all the commotion. "You-all can't go anywhere just yet. Your urine samples came back with questionable results, which means that you're going to have to give another sample."

"Holy shit," Johnson snorted when the Sergeant was out of hearing range. "Wouldn't you just know it? Gutmann's pee flunked the test."

"Yeah. And more importantly, *mine, yours, and Lee's did too*," I responded disappointedly.

That same afternoon we again submitted urine samples. This time each of us used his own and the results were okay. The next day, a day later than planned, we walked through the main gate at Fort Jackson for the last time. Gutmann may still be there for all I know. Sadly, I never saw him or any of the other fellows again.

EPILOGUE

Back in Raleigh, I learned that I had been "provisionally" accepted to enter the fall semester at Campbell College. The "provision" was that I pass four three-semester-hour classes (One English, two World History, and Psychology) at North Carolina State College during the summer, with at least a C average. I made Bs in all four, and began to plan for my full-time college education, which started in September. The first thing I did was take money that I'd saved from active duty and bought a used British Triumph TR3. It was white, with Kentucky blue fold-down top and side panels. Next, I went to Nowell's clothing store in Cameron Village and purchased Bass Weejun loafers, two pairs of Madras shorts, and two Gant shirts—one light blue and the other yellow. Evolving from military to preppy in short order, I was ready to hit the books.

As the years passed by, and far removed from campuses of my college days, I have often thought of my old buddies at Fort Jackson. And I still do. This is particularly true when I see someone in an Army uniform, or watch a war movie. I probably think of Pvt. Plank more than the rest. I have been somewhat unkind to him in this text. I would be surprised if he had not gone on to Airborne training after leaving Fort Jackson, and then become a Ranger or Green Beret. I think too that he probably went over to Viet Nam, and if I had to wager I'd bet he died there. He was Gung Ho you remember, and would've been in the heat of the action and leading assaults as he did at Jackson. Some of my other Fort Jackson pals, ones I would've never expected to go into battle, probably served there with distinction as well.

I was honorably discharged from the Service in 1968. Today, looking back, I realize that our time together at Fort Jackson was short but we were forever bound by our common experience. We were molded into men during that fall, winter, and spring as we marched through coal smoke haze and jogged through the sand hills of South Carolina. It wouldn't be such a bad idea for most young men and women in this country to have to do the same thing. It sure worked for me, and I'll forever be grateful for it.

ABOUT THE AUTHOR

Chuck Manooch was raised in Raleigh where he attended public schools, and graduated from Needham Broughton High School in 1961. He also graduated from Campbell College in 1966 and North Carolina State University where he obtained a M.S. in 1972 and Ph.D. in 1975, both in Zoology. From 1972 through 2002 Chuck was employed as a marine scientist at the NOAA Beaufort, NC Laboratory. He has served as an Adjunct Professor at North Carolina State University, East Carolina University, University of North Carolina at Wilmington, and Rutgers University. As an Adjunct Professor he has served on more than thirty graduate student committees. During his career he has published more than one hundred scientific papers, and authored three other books: *Spring Comes to the Roanoke,* in 1979, *Fisherman's Guide to the Fishes of the Southeastern United States*, first published in 1984, and *Growing up an Old Raleigh Boy,* first published in 2005. Chuck volunteers at his church and at the NC Division of Marine Fisheries. He is a member of the Beaufort Ole Towne Rotary Club, the Carolina Country Club, and especially enjoys being with his buddies at the Long Hope Hunting Club. Chuck and his wife Carol reside in Morehead City, NC.